SpringerBriefs in Plant Science

For further volumes:
http://www.springer.com/series/10080

M. Swapna · Sangeeta Srivastava

Molecular Marker Applications for Improving Sugar Content in Sugarcane

 Springer

M. Swapna
Division of Crop Improvement
Indian Institute of Sugarcane Research
Lucknow 226002
Uttar Pradesh
India

Sangeeta Srivastava
Division of Crop Improvement
Indian Institute of Sugarcane Research
Lucknow 226002
Uttar Pradesh
India

ISSN 2192-1229
ISBN 978-1-4614-2256-3
DOI 10.1007/978-1-4614-2257-0
Springer New York Heidelberg Dordrecht London

e-ISSN 2192-1210
e-ISBN 978-1-4614-2257-0

Library of Congress Control Number: 2012931948

Printed on acid-free paper

Springer is part of Springer Science+Business Media (www.springer.com)

Foreword

As a C_4 plant, sugarcane has very efficient system for carbohydrate metabolism through photosynthesis and sugar accumulation. Crop improvement efforts have concentrated mainly on improving quality traits, mainly sugar content. This being a complex trait, involves a large number of target genes in the metabolic pathway. The complex polyploid nature of the crop makes it more difficult to pin point the key players in this complex pathway. Despite its importance, little is known about the exact mechanism of sucrose accumulation and its regulation in sugarcane. Many enzymes have been proposed to have a key role in determining the ultimate sucrose content in sugarcane. Especially in a crop like sugarcane where the classical techniques are of limited help in elucidating various genetic complexities, molecular techniques can be of help in throwing some light on the grey areas. Molecular marker strategies will help in understanding some aspects of sucrose metabolism and its regulation in this crop, thus complementing the ongoing crop improvement programmes.

The review seeks to look into the crop improvement programmes in brief, in this crop, gradually resulting in the biotechnological interventions. The initial studies on molecular markers with respect to quality attributes, mainly sugar content, like diversity studies, marker identification, mapping strategies and other applications leading to functional genomics and the impact of these techniques in improving the sugar content, directly or indirectly have been dealt with. The possibility of diverse roles being played by some of the genes calls for more detailed analyses to study their possible role in sugar accumulation. Comparative genomics, which has an important role in the genomic analysis in this crop is another aspect reviewed. The complexities Associated with the crop make it a difficult candidate for molecular studies too, compared to other crops. The possible implications of the polyploidy and other peculiarities of this crop will help us to be cautious in our approach towards molecular marker applications in this crop, even though the rapid advancements in this field may be of help in overcoming many of these difficulties. Thus this book will serve as a useful guide to researchers who are engaged in molecular genetic studies related to quality attributes in sugarcane.

M. Swapna
Sangeeta Srivastava

Contents

Molecular Marker Applications for Improving Sugar Content in Sugarcane

1 Introduction

Sugarcane is an important crop which plays a substantial role in the world economy. About 75% of the world sugar is produced from sugarcane and the rest is from sugarbeet. The mature stalks of sugarcane contain juice with sucrose content as low as 8–9% to 18–19%. Sugarcane is unique in the sense that storage of sucrose in the storage parenchymatous tissue takes place at a very high concentration, as against starch storage in most other higher plants. Sugarbeet and sweet sorghum are other important crops that store photosynthetic assimilates in the form of sucrose. Apart from the traditional use as a source of sugar, sugarcane is fast becoming a source for ethanol and biomass production as an alternative energy source. The residue after sugar extraction from the sugarcane stalk, i.e., bagasse, is a used for generating electricity. Direct utilization of bagasse for bioethanol production has also been facilitated by novel fermentation technologies. Even though the concept of sugarcane as a bioenergy crop is fast picking up, the use of sugarcane as a primary source of sugar remains the top priority till now. Sucrose is considered to be the major sugar, making upto almost 60% of the dry weight of sugarcane culm, with glucose and fructose present at lower concentrations. Thus sucrose accumulation, its retention in the stalks and regulation of the process assumes great significance.

Sugarcane is a complex polyploid belonging to the subtribe Saccharinae of tribe Andropogoneae. The genera *Saccharum*, *Erianthus* (sect. *Rimpidium*), *Miscanthus* (sect. *Diantra*), *Sclerostachya* and *Narenga* constitute a closely related inbreeding group referred to as "*Saccharum complex*" (Mukherjee 1957). The genus *Saccharum* comprises of six species—*Saccharum officinarum*, *S. barberi*, *S. sinense*, *S. edule*, *S. robustum* and *S. spontaneum*—the first four are cultivated and the last two grow wild in nature (Daniels and Roach 1987). Modern sugarcane varieties are derived from the interspecific hybridization involving *S. officinarum*, *S. spontaneum* and *S. barberi* with contributions from

M. Swapna and S. Srivastava, *Molecular Marker Applications for Improving Sugar Content in Sugarcane*, SpringerBriefs in Plant Science, DOI: 10.1007/978-1-4614-2257-0_1, © The Author(s) 2012

other species also. For the successful utilization of any crop in breeding, the interrelationship among the different members of the species/genus/related genera has to be understood. Several workers have suggested various groupings among the *Saccharum* complex and related genera with modifications from time to time (Dutt and Rao 1951; Daniels and Roach 1987).

Sugarcane improvement, from selection of existing variation in pre-historic time to the current bi/multi-parental crossing and subsequent use of non-conventional techniques, has concentrated mostly on improving the yield and sugar content. Due to the complexities associated with this crop, not many investigations were made regarding the genetics and inheritance of important traits. Since most of the economically important characters are quantitative in nature, the quantitative inheritance has been studied. Hogarth (1968) and Brown et al. (1968) have reviewed some of the problems associated with quantitative genetic studies in this crop. The general assumptions for any quantitative genetic analysis may not be met in the case of sugarcane. Even then, the developments in the field of genetics and breeding—both at classical and molecular level—has helped us to make a quantum leap in the area of sugarcane improvement.

2 Sugarcane Cytology

The cytology of the crop has been studied in great detail by earlier workers (Bremer 1961a, b, c, d; Price 1964, 1965, 1968, 1969). It has been described as a complex autopolyploid with many cytogenetic complexities like high chromosome number, variability in chromosome number, *en masse* elimination, female restitution in interspecific or even in species X cultivar, crosses etc. Mostly, the chromosome number in various species and cultivated clones were studied during the early years, with the numbers varying from 80–120. In general, normal bivalent formation, low to fairly high frequencies of univalents, and other irregularities like laggards, bridges and spindle abnormalities have been reported in different species level clones and species hybrids. The cultivated clones mostly have a range of chromosome number between 100 and 130 (Sreenivasan et al. 1987). Chromosomal studies conducted earlier were not confined to *Saccharum* species clones or cultivated hybrids. Interspecific hybrids, intergeneric hybrids, crosses with related genera like *Erianthus* and *Miscanthus*, *Sorghum*, *Zea mays* etc., were also studied in detail by earlier researchers (Jagathesan and Ramadevi 1969; Jagathesan and Sreenivasan 1967, 1971; Jalaja 1983; Janaki Ammal 1938a, b; Janaki Ammal et al. 1972). Chromosomal irregularities were reported in all the above material.

An important peculiarity in sugarcane is the 2n + n transmission that has been observed in interspecific crosses. Interspecific hybrids, particularly involving *S. officinarum* × *S. spontaneum* with *S.officinarum* as the female parent, have a triploid chromosome number relative to the parents (Bremer 1923, 1924, 1925, 1961a, b, 1962; Dutt and Rao 1933). This was attributed to the transmission of double the haploid set (2n) of chromosomes by the female parent and haploid set (n) of chromosomes by the male parent. It has also been reported that this increase in chromosome number continues only up to the second back cross (Bremer 1961b, c, 1962). 2n+n transmission of chromosomes has been reported in *S. officinarum* X cultivar crosses also (Piperidis et al. 2010) indicating that this phenomenon is not confined to interspecific crosses alone. Even though several mechanisms have been postulated to explain this phenomenon the exact process has not been unambiguously established. Unreduced egg cells, endoduplication, post-meiotic fusion, endomitosis, differential survival of gametes, selective zygotic failure etc., are some of the possible reasons suggested (Sreenivasan et al. 1987; Bhatt and Gill 1985). These cytological peculiarities invariably effect the pattern of meiotic chromosomal transmission and thereby, the inheritance of important economic traits. Needless to say, these will have an effect on the efficiency of techniques like molecular marker applications.

A comparison of the initial advancement in the field of biotechnological interventions in other crops like rice, wheat, maize and in sugarcane brings out the fact that the initial progress in sugarcane was not as quick as in other crops. It is here that conventional cytological tools assume significance. If we look into the cytogenetic studies in these crops, availability of cytogenetic tools like monosomics, nullisomics, addition-substitution lines, transposons etc. have had important roles in conventional gene location and mapping studies. In rice a complete series of monosomic alien addition lines (MAALs) were produced (Jena and Khush 1989), each with a complete set of chromosomal complement from *Oryza sativa* and a single chromosome from *O.officinalis*. Such lines have been useful in transferring important traits to cultivated rice and also to study the gene/chromosomal effects. In maize, tomato and barley extensive studies were undertaken to assign linkage groups to specific chromosomes using trisomics (McClintock and Hill 1931; Rick and Barton 1954; Rick et al. 1964; Tsuchiya 1959; Riley et al. 1968). Trisomics have also been used in locating genes on specific chromosomes and for physical mapping (Tsuchiya 1991). Monosomics and nullisomics have been used in locating genes to specific chromosomes and chromosome arms in wheat. Transposon-tagging has helped in initial genetic studies in maize. Thus a lot of work was done using these conventional cytogenetic tools in chromosome mapping, physical mapping etc. This had given the workers of these crops a starting point from where, molecular tools had taken over. In the case of sugarcane, the large genome and the numerical complexities, along with the small size of individual chromosomes had been a hindrance to the development and successful utilization of such cytogenetic tools. The buffering capacity of this genome would also have been a drawback to study any gene/chromosomal effects in monosomics, nullisomics or substitution lines. Thus no such cytogenetics based gene location or chromosome mapping was carried out in this crop during the early days. Researchers employing the molecular tools did not have much initial data to rely upon, thus slowing down the pace of molecular studies in this crop at the initial stages. But with the advent of advanced molecular tools, the pace in this crop has also picked up.

3 Breeding in Sugarcane for Quality Attributes

As mentioned before, crop improvement efforts have mainly concentrated on improving the sugar yield. Final sugar yield involves many component traits like stalk length and stalk girth, sugar content and cane yield. Thus sugar yield can be improved by improving the sugar content or the cane yield. Selection criteria in sugarcane include commercial cane sugar (CCS) apart from cane yield. The sugar content in the commercial crop has reached a plateau with not much improvement being brought about by conventional practices. Thus, many a times, cane yield forms a major criteria for selection. Since this is an industrial crop, the sugar factories and mills which procure the canes from sugarcane growers, play a major role in the sugarcane development scenario. In many countries including India, the sugarcane growers are being paid on the basis of cane yield rather than by the final sugar content (even though there are incentives for growing an early maturing high sugar variety). Thus many a times, the emphasis for breeding shifts to a better yielder along with the sugar content. This might be one reason due to which the molecular interventions for increasing sugar content per se has not been put to practice to a greater extent in this crop.

A successful breeding programme necessitates proper flowering of the parental lines. Suitability of different climatic zones for sugarcane flowering varies. For example in India, Coimbatore at southern part of the country is best suited for flowering. Here also, a lot of variation exists among different clones with respect to flowering and extent of flowering. Temporal variation can also be observed for the same genotype. This, to some extent, creates difficulty in repeating specific crosses during subsequent years. The flowers are bisexual and depending on the pollen fertility the individual varieties/clones are designated as male or female. No emasculation is carried out in artificial crossing programmes and thus, unless a genotype is male sterile with zero pollen fertility, there may be chances of selfs even among progenies of controlled crosses. Some desired genotypes may not flower at all during a particular year or the flowering of desired parents may not synchronize at a given time. Thus the sugarcane breeding programmes have been more of a "hit and miss" affair. All these peculiarities in the reproductive biology of this crop determine to what extent the application of molecular marker techniques (and for that the matter, that of other biotechnological tools) are successful in this crop.

As in any other crop the conventional methods of selection, hybridization etc. have been the method of choice for breeding in sugarcane also. Collection, maintenance and proper evaluation of germplasm is an important prerequisite for success in breeding. Traditional breeding starts with the selection of parental material from the source population. For each breeding programme the selection of parental material depends on the aim of the programme. Since the commercial breeding activities invariably emphasizes on high sugar varieties, the parental material for commercial breeding programme essentially involves utilization of high sugar parents. Hybridization follows, with the crosses consisting of biparental crosses, polycrosses or general crosses (where the pollen from many unknown parents can pollinate a single female parent). Selfing too forms a part of varietal development strategy, as enough variation can be obtained in selfed progeny also. Thus this can also serve as a source population for elite clone selection. In India as mentioned before, Coimbatore situated in the tropical region (77°E longitude and 11°N latitude) is endowed with suitable climatic conditions for flowering. So most of the hybridization programmes are carried out at the National Hybridization Garden (NHG) situated at Sugarcane Breeding Institute, Coimbatore, established under the All India Coordinated Research Project (AICRP) on Sugarcane (even though some individual research centres make crosses on a small scale at some of their own facilties. After seed set the fluff is collected, dried appropriately and is sent to the different research centres for further studies. The fluff obtained after crossing are sown in glasshouse or mist chambers. The seedlings can be transplanted to a secondary nursery or to individual polybags at about 2 months stage. Under proper conditions the seedlings are transplanted to the field. Since each clump has its own unique genetic constitution, individual clumps can be considered as separate entity. Individual seedlings are subjected to hand refractometer brix (HR Brix) observations to identify the high sugar clones. This is followed by selection of suitable clones. These are advanced further and after a series of vegetative propagation cycles the superior clones are released as elite varieties for commercial cultivation.

The involvement of high sugar parents in hybridization invariably calls for some pre-breeding strategies which forms an important component of crop improvement programmes. A pre-breeding programme for high sugar parental development (and also for other traits) exists in most of the crop improvement activities. Pre-breeding for high sugar involves crossing among high sugar elite clones and indigenous as well as exotic parents and further selection. The idea is essentially to introgress new gene combinations in the progeny that can eventually result in high sugar early maturing clones. For this a recurrent selection cycle for high sugar can be employed. The initial cycle of crossing and selection will give rise to a set of high sugar parents, which in turn can be intercrossed among themselves. Alternately, this can also be an outcome of the regular varietal development programmes, For example, elite clones which have high sugar content but do not meet the expectation of a commercial variety due to disease and pest incidence, low yield etc., can be used as parental lines after studying their reproductive behaviour. One such programme at India has yielded a large number of high sugar breeding stocks which have been included in the National Hybridization Garden facility for use by other researchers in their breeding programmes. Selection history and contribution of *S. spontaneum* to the clone's pedigree influence adaptation to sugar yield and its components as proposed by Srivastava et al. (1994). In a study involving a random sample of 64 clones from the germplasm collection of CSR, Australia, and their bi-parental progenies, (6 each, from each of the 32 full-sib families) a strong association, was recorded between the level of expression of characters including those for sugar, with the minimum number of backcrosses to *S. spontaneum* in the pedigree of the clones. Importance of *S. spontaneum* component was suggested by Reffay et al. (2005) also in his molecular marker studies. Roach (1989) had argued that clones which had undergone less nobilization would be better suited to stress and other unfavourable conditions.

High correlation of brix with sucrose content was reported very early (Hebert and Henderson 1959) with the suggestion that selection for high sucrose is possible at single stool stage itself. Sugar accumulation and final sugar content have been reported to be quantitative characters with more than one gene affecting the trait. Most of the variation was attributed to additive effects (Hogarth et al. 1981). The role of female parent in the inheritance of brix in early stages has been reported to be more, compared to that of the male parent by several workers, even though the role of male parent cannot be ignored (Hsu et al. 1995; Shanthi et al. 2005).

4 Sugar Metabolism in Sugarcane

Sugarcane is a C_4 plant which very efficiently transforms carbohydrates into sugars. The crop is unique in the sense that storage of sucrose in the storage parenchymatous tissues takes place at a very high concentration as against starch storage in other higher plants. Sucrose is considered to be the major sugar making up almost 60% of the dry weight of sugarcane culm, with glucose and fructose present at lower concentrations. Thus sucrose accumulation, its retention in the tissues and the regulation of the entire process are important. As in any other crop, a large number of enzymes are involved in the sugar metabolism in this crop. Sugarcane is a crop which grows in tropical and sub tropical conditions. For example in India, the sub-tropical region comprises of a major sugarcane growing belt. Here the spring planting is done during February–March when the temperatures are not very high. The germination is followed by the tillering and grand growth phase when the temperatures are very high (40–45°C). This continues till August–September coinciding with the monsoon season when the biomass accumulation takes place. By the onset of winters the temperature decreases and carbon partitioning into sucrose takes place resulting in sucrose accumulation. Autumn planting is also adopted by some growers where the planting takes place during October, at the onset of winter and the crop is in the field for 12–16 months. Thus the crop passes through a gamut of environmental conditions in the course of its life cycle. In tropical belt, sugarcane grows very well even though the temperature conditions are significantly different. Here the crop enjoys an extended growing period the resulting in good yield. Thus several factors can be identified that may influence sugar accumulation and ripening of sugarcane in these different climatic conditions.

Let us have a brief overview of the sucrose metabolism pathway that is well-known in sugarcane. Sugarcane is a C_4 plant where most of the carbon dioxide is initially fixed in a 4-carbon form—malic and aspartic acid (Kortschak et al. 1965; Hatch 1977). The occurrence of Kranz anatomy in the sugarcane leaf cells is specific to this C_4 cycle. In monocots where C_4 photosynthesis occurs the leaf anatomy is peculiar. The vascular bundles are invariably covered with one or two distinct layers of thick-walled bundle sheath cells, thus separating them from the mesophyll cells. This concentric arrangement of bundle sheath cells gives rise to the Kranz anatomy (Kranz—wreath like, in German) (Laetsch 1974). These bundle sheath cells have thicker walls than that in C_3 plants (where the bundle sheath cells are themselves less prominent), contain more chloroplasts, mito-chondria and other organelles and smaller vacuoles. It has been observed that there is a clear cut spatial demarcation between the site of C_4 acid formation and that of sucrose-starch production. While the first reaction takes place in the mesophyll cells 3-PGA, sucrose and starch are present in the bundle sheath cells. Thus the complete Calvin Cycle occurs in the bundle sheath cells and the C_4 acid formation occurs in the mesophyll cells. The initial developmental pathway during leaf formation and development resulting in this peculiar leaf architecture thus favours the C_4 type of photosynthesis. How great a role does this peculiar leaf anatomy *per se* have in the high sugar accumulation capacity exhibited by these C_4 plants needs to be studied so that the genes responsible for this architecture can also be effectively put to use in future breeding programmes.

The C_4 plants are found to be more efficient especially under high temperature conditions. The problem of stomatal closing and reduced CO_2 availability during high temperature encountered by the C_3 plants is not observed in C_4 plants. There is an effective pumping of CO_2 into the bundle sheaths cells thus doing away with the possibility of CO_2 being a limiting factor in this group of plants. Also, the losses due to photorespiration are practically nil in C_4 plants compared to C_3 plants. All these factors together contribute to an increased efficiency in C_4 plants (Salisbury and Rose 1992). The potential of these plants is best exploited under tropical conditions, even though these perform well under sub-tropical/temperate conditions also.

As can be visualized, several factors are responsible for the ultimate sugar accumulation in a genotype. The total amount of sucrose resulting from pho-tosynthesis also forms the source for signalling for modulation processes, transport and storage, flowering induction, differentiation processes etc. Thus apart from the genes directly responsible for sucrose synthesis and/or its breakdown, the genes involved in these metabolic pathways will have a role in determining the final sugar content of a variety (Ming et al. 2006). Thus there can be a large number of genes that are linked with final sugar content. Needless to say, the variations observed with respect to these aspects, among the sugar-cane clones may also have to be studied to explain the differences in sucrose accumulation capacity observed among the different clones. Ming et al. (2006) have reviewed the possible pathways involved in sugar accumulation. The important pathways suggested may be depicted as those occurring at different points throughout the plant.

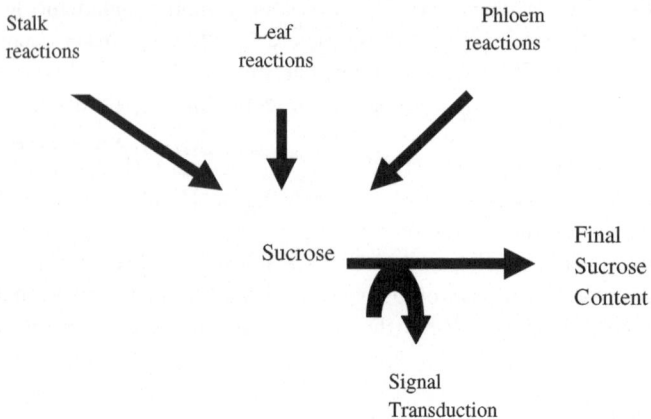

The photosynthesis pathway, sucrose metabolism and carbon partitioning taking place in leaf, sucrose transport and its translocation in the phloem tissues, and partitioning and remobilization of sucrose synthesized in the stalk tissues along with the various signalling functions involving sucrose, determine the final sucrose content in the variety. There are enzymes which are involved in the photosynthetic cycle. The enzymes responsible for sucrose synthesis and break down also play significant role in the final sugar content. The bio-synthesis of sucrose is catalyzed by Sucrose Phosphate Synthase (SPS) and Sucrose Phosphate Phosphatase (SPPase). Sucrose Synthase (SuSy) is another enzyme involved in carbohydrate metabolism. In monocots two non-allelic forms of the enzyme—SuSy1 and SuSy2 are found to be involved [a third form SuSy3 being reported in rice by Huang et al. (1996)]. Acid and neutral invertases are another set of enzymes that play a major role in sucrose brake down at different stages of maturity. Once sucrose is synthesized, this needs to be transported to the different storage sites, with different transporters responsible for this. Other probable candidates are the genes responsible for cell membrane permeability, genes related to osmotic balance, carbon partitioning etc. Apart from these metabolic pathways the genotypic and environmental effects also play a significant role in the final sucrose content. The timing of maturation adds another dimension to the process.

Overall, the entire network of sucrose synthesis, accumulation, storage and retention is a complex mechanism where several metabolic pathways interact with each other. It is therefore not surprising that the conventional methods of genetics and related branches have been not able to fully elucidate this complex metabolic pathway. It is in this context that the role of the modern tools of molecular biology assumes significance in elucidating the sugar accumulation mechanism, thus complementing the conventional breeding programmes.

5 Molecular Markers and Their Use in Sugarcane

Before going into the details of molecular marker applications, let us have a very brief overview of the molecular markers commonly used for different applications in sugarcane.

Any character or trait that can differentiate among individuals and that are accurately heritable can be considered as a marker. These can be either visual markers or biochemical markers like enzymes or molecular markers. Molecular markers can be molecular cytogenetic markers based on variations in chromosomal segments or DNA based markers based on differences at DNA sequence level. The pre-dominant markers used during earlier days were morphological and biochemical markers. With advent of recombinant DNA technology the use of molecular markers especially DNA based markers became more common. DNA based markers are small regions of DNA showing sequence polymorphism in different individuals that are heritable. Some of the salient advantages that are associated with DNA based markers are:

- Abundance in the genome
- Extensive genomic coverage
- High levels of polymorphism
- Independent of developmental stage
- No tissue specificity
- Less environmental sensitivity (though recent findings with respect to QTL analysis points towards the possibility of environmental dependence at least in some cases).

A very brief description of these markers is presented here (reviewed in Prasanna 2002) though it may be very preliminary to many readers.

5.1 Different Types of Molecular Markers

Restriction Fragment Length Polymorphism (RFLP). RFLP markers are based on polymorphism generated by restriction enzymes and this is an application of southern hybridization (Botstein et al. 1980). These markers are defined by specific probe-enzyme combinations. A typical assay would involve the following procedures:

- Digestion of plant genomic DNA with restriction endonucleases
- Separation of generated fragments by electrophoresis
- Southern blotting and southern hybridization to expose the membranes to suitable probes
- Autoradiography/Chemiluminescence to detect polymorphism

Polymorphism is detected due to different sizes of the restriction fragments in different genotypes to which the probe may be homologous.

PCR based markers. These markers make use of the Polymerase Chain Reaction (PCR) where small DNA fragments are amplified many times in a differential manner. This procedure involves the denaturation of DNA strands, annealing of primers to the complementary sequence in the DNA strands and extension of the primes to form short DNA fragments of 200 bp or even up to 3–4 kb, using a thermal cycler. The polymorphism arises due to differential amplification of DNA sequences in different genotypes and can be visualized by electrophoresis or other suitable non-gel based methods.

The PCR based markers may involve the use of random primers as in Random Amplified Polymorphic DNA (RAPD) (Williams et al. 1990) where the target DNA is amplified many times using random primers. Here there is no pre-requisite of knowledge of gene sequence information, but there is the disadvantage of lack of reproducibility as the assay is sensitive to experimental conditions.

Specific sequences or sequences from known genes can also be used as in **Simple Sequence Repeats (SSR) or Sequence Tagged Sites (STS) or microsatellite markers**. Microsatellite markers are short repetitive sequences which help in identifying the polymorphism based on Polymerase Chain Reaction (Gupta et al. 1996; Powell et al. 1996). Here sequence specific primers are used instead of random primers. Thus the repeatability and reliability are more in this case. These are powerful tools for genotype differentiation, genetic diversity analysis, purity evaluation of seeds, mapping studies, marker assisted selection etc.

Amplified Fragment Length Polymorphism (AFLP). AFLP combines both restriction and amplification of DNA using various combinations of restriction enzymes and primers and the variability is visualized by autoradiography (Vos et al. 1995). This involves

- Digestion of genomic DNA with restriction enzymes-a rare cutter and a frequent cutter
- Ligation of adapters to the cut end
- Pre-selective and selective amplification
- Separation of the fragments by electrophoresis
- Visualization by autoradiography

Single Nucleotide Polymorphism (SNP). SNP are point mutations in which one nucleotide is substituted at a particular locus. They represent an inexhaustible source of polymorphism which are useful in high resolution mapping studies. These can put to use where the genomics is well advanced.

Apart from these markers, several markers based on these marker systems have been developed which have also been tried in sugarcane. **Expressed Sequence Tag (EST) based markers** utilize the expressed portion of the genome or the cDNAs (Sasaki et al. 1994; Yamamoto and Sasaki 1997; Wang and Bowen 1998; Cho et al. 2000; Scott et al. 2000). EST sequences represent real-function genes and thus are more useful as genetic markers. These provide useful targets for incorporation into genetic maps. The increased transferability of these sequences across genera due to their presence in the conserved regions is another important advantage of these markers. Especially in a crop like sugarcane with a complex genome, where the use of classical tools are of limited advantage, EST derived markers are useful candidates for accessing genetic information. A large number of ESTs have already been reported in sugarcane. The sugarcane EST project SUCEST has built a database containing 2,38,000 ESTs from 26 cDNA libraries constructed from several organs and tissues sampled at different intervals (http://sucest.lad.ic.unicamp.br/en/). The EST database in the public domain also serve as a readily available inexpensive source of microsatellite markers. **Single Strand Conformation Polymorphism (SSCP)** from genomic sequences as well as ESTs (Swapna et al. 2011a) has been used in sugarcane also, as in other crops (Fukuoka et al. 1994). Here the amplified products are converted into single strands and electrophoresed. Polymorphism arising out of conformational changes in the single strand are visualized in this case. The exact nature of sequence differences that result in formation of different secondary structures as reflected in the shift in the mobility in the gel remains unclear. This could be due to sequence rearrangements, single nucleotide insertion-deletions or substitutions or more than one of these changes as has been reported in other plants (Fukuoka et al. 1994; Bodenes et al. 1996). The exact molecular nature of these variations will be understood after cloning and sequencing of the individual conformers. **Targeted Region Amplified Polymorphism (TRAP)** is a PCR based marker system where an EST sequence is used to design primers along with an arbitrary sequence (Alwala et al. 2006a, b). A fixed primer is designed from an EST sequence and an arbitrary primer of the same length with an AT or GC rich motif (to anneal with an intron or exon respectively) is designed. **Sequence Related Amplified Polymorphism (SRAP)** has also been used in sugarcane for various purposes like mapping studies (Alwala et al. 2009). Conserved Intron Scanning Primers (CISP) is another marker system based on the conserved sequences that has been put to use in sugarcane also (Khan et al. 2011).

In sugarcane, all these molecular marker systems have been put to use to assess the naturally occurring genetic variability, to construct genome maps and to tag genes for economically important traits and for comparative and functional genomics studies. (D'Hont et al. 1993, 1994; Lu et al. 1994; Nair et al. 1999; Ming et al. 1998, 2001, 2002; Aitken et al. 2005, 2006, 2008; Pinto et al. 2004; Alwala et al. 2006a, b, 2008, 2009; Parida et al. 2009, 2010; Swapna et al. 2011a, b; Singh et al. 2008, 2011).

5.1.1 Other Marker Systems

Ribosomal DNA and RNA (rDNA and rRNA) and other conserved sequences have been used in sugarcane for phylogenetic and diversity studies (Glaszmann et al. 1990; D'Hont et al. 1998). Molecular cytogenetic techniques like Genomic and Flourescent in situ hybridization (GISH, FISH) have also been used on a large scale in sugarcane genome studies especially for genome analysis and related studies (D'Hont et al. 1995, 1996). These investigations have helped in determination of basic chromosome numbers of *Saccharum* spp. clones and hybrid, genome characterization in modern cultivars and in related genera, understanding the pairing behaviour etc.

Let us have a look at the various broad areas of molecular marker application. As is common knowledge, the success of any crop improvement programme depends on the variability present in a population. Recent studies reveal that variation exists to a great extent in the naturally existing populations in sugarcane. This variation has to be exploited for its effective utilization in plant breeding programmes. Thus germplasm evaluation and screening are major areas where molecular markers can be utilized through diversity studies, fingerprinting etc. Another area of molecular application involves the hybridization carried out among desirable parental clones. Here several aspects come into picture. Selection of parents, development of suitable parents i.e., pre-breeding, crossing and further selection in different generations, back crossing to introgress desirable genes/loci into a well-adapted variety, gene introgression for genetic base broadening, etc., are some of the areas where molecular markers can be used. Thus exploiting the existing variability, as well as creation of new variability are important crop improvement activities where molecular markers can be put to use. Molecular markers are useful in transgenic research. If the loci linked to the marker are identified for any particular trait, the sequence as such may be useful for transformation studies. On the other hand, once transgenics for a particular trait are developed, efficient markers if identified can be used for screening the transformants. The scope of development of new molecular markers on the basis of information generated from transgenics is also a possibility. Testing for distinctness, uniformity and stability (DUS testing) in a newly developed variety, genetic purity testing to identify admixtures are other significant areas where molecular markers play a role (Santhy et al. 2000; Prasada Rao et al. 2001). Differentiation of varieties based on expressed genes assumes significance in the light of UPOV initiatives aiming at establishing the distinctness, uniformity and stability (DUS) using molecular markers. As outlined in the Article I of the 1991 UPOV Convention (UPOV 1992), a plant variety is defined by the expression of characteristics and is distinguished from any other variety by the expression of at least one of the characteristics. Distinctness of a variety established by DNA markers based on non-coding regions of the genome therefore may not meet the above requirement. Use of markers derived from the expressed genes is therefore more appropriate, particularly for a plant species with a large polyploid genome. Comparative genomics and functional genomics are other important areas where information using molecular markers can be put to use.

5.2 Molecular Markers in Sugarcane: Applications

5.2.1 Germplasm Characterization and Diversity Studies for Sugar

Different *Saccharum* spp. clones and commercial hybrids show wide variation with respect to the sucrose accumulation potential. Even though environmental conditions may influence sugar accumulation and ripening to some extent genotypic differences are quite prominent for this trait. These differences have been exploited by sugarcane researchers for selecting high sugar progenies from large populations. Molecular marker studies for germplasm characterization has been carried out by researchers from early days. Utilization of different types of markers like RFLP, RAPD, SSRs, AFLP etc. have helped in understanding the diversity present in different germplasm all over the world. Since most of the diversity studies include the commercial high sugar varieties, the general diversity studies in sugarcane will indirectly aim at studying the variation with respect to sugar content at the molecular level. These have also been reviewed here.

When we talk about sugar accumulation in sugarcane, the very early works by Hatch and his group (Hatch et al. 1963; Hatch and Glasziou 1963; Sacher et al. 1963) invariably finds a mention, even though strictly speaking their work does not involve any molecular markers. Through their series of experiments using tissue slices they had tried to identify some of the enzymes involved in sugar metabolism in this crop. The salient findings include the proposal of a cyclic scheme for sugar accumulation involving three distinct cell compartments viz., the outer space, the metabolic compartment and the storage compartment. These workers identified the presence of acid and neutral invertases in the immature and mature sugarcane stems respectively. It was also suggested that the acid-invertase mediated inversion taking place in the outer space, is an integral step in sucrose accumulation and may even act as a rate limiting step. A schematic representation was suggested for sucrose accumulation (Sacher et al. 1963).

In essence this was a preliminary assessment of the variation exhibited within the stalks in different varieties of sugarcane with respect to the sugar related enzyme activities. From these differences they had formulated a proposed cycle for sugar accumulation. Lingle (1996, 1997, 1999) had also studied the differences with respect to sugar accumulation and related biochemical processes in different sugarcane genotypes to come to certain conclusions regarding the sugar accumulation process in this crop. They postulated that acid invertase suppresses sucrose accumulation in elongating internode (Lingle 1997) while this role may be carried out by sucrose synthase in fully elongated internodes (Lingle 1996). Rate of sugar accumulation was found to be faster in late developing internodes. The role of sucrose phosphate synthase and acid invertase in determining sucrose concentration during maturation in sugarcane internodes was emphasized by these workers (Lingle 1999).

Use of biochemical markers like isozymes was reported for genetic diversity studies in sugarcane (Glazsmann et al. 1989). Later the work on DNA based markers started gaining momentum in this crop. Initial studies were dominated by RFLP markers. With the advent of much simpler PCR based markers random markers were used followed by other markers like SSRs, AFLP, SNP, SSCP based markers and TRAP markers. Since many of the works involved the use of commercial cultivars/varieties from different regions, it is assumed that indirectly these studies lead to diversity with respect to sugar content and sugar yield, even though the markers for sugar content may not have been specified.

The initial molecular markers for studying germplasm and diversity comprised of nuclear ribosomal DNA markers (rDNA) (Glaszmann et al. 1990). Restriction studies using mitochondrial (D'Hont et al. 1993) and chloroplast DNA (Sobral et al. 1994) indicated that *S. spontaneum* had a distinctly different restriction pattern. Nuclear RFLPs analysis of germplasm diversity (Lu et al. 1994; Burnquist et al. 1992; Jannoo et al. 1999) exhibited a high level of heterozygosity among cultivated clones. AFLP markers were also used to study diversity and to characterize germplasm in sugarcane (Lima et al. 2002; Selvi et al. 2005, 2006). But the most widely used markers till date are the microsatellite based markers-either genomic or EST (Cordeiro et al. 1999, 2000, 2001; Pan et al. 2004; Pan 2006; Hemaprabha et al. 2006; Oliveira et al. 2007; Selvi et al. 2003; Parida et al. 2009, 2010; Swapna et al. 2011a; Singh et al. 2011). The different reports have revealed different degrees of polymorphism among the sugarcane clones depending on the plant material used and the marker system studied. In general the species level clones exhibited a larger degree of variability with the molecular markers. Limited levels of variability have also been reported among cultivated clones (Harvey et al. 1994; Nair et al. 2002) thus pointing towards a very limited capture of naturally existing variation in the cultivated clones.

The moderately high genetic diversity observed among the cultivated varieties commonly used in crossing programmes (Hemaprabha et al. 2006) opens up an excellent opportunity for their utilization in commercial breeding. Since intercrossing of elite hybrids is a main component in varietal development, such studies based on molecular marker information can be helpful in selecting diverse parental combinations giving maximum diversity with respect to sugar content. This can be an excellent strategy to concentrate different sucrose genes/alleles, facilitating the build up of breeding stocks for sugar content, arising out of diversity and similarity observed among closely related commercial varieties (Hemaprabha et al. 2006). This points towards the scope for application of molecular markers not only in varietal breeding, but also in pre-breeding, there by aiding the pooling of sucrose genes through inbreeding, as suggested by Stevenson (1965).

The use of EST derived microsatellites for diversity studies have also revealed the underlying variation existing in the germplasm. Through these markers, the functional diversity has also been revealed among the different clones. Such studies involving genomic and EST derived SSRs, using *Saccharum* spp. clones, related genera and commercial hybrids demonstrate the contrast among the two classes of microsatellites in their potential to reveal diversity (Cordeiro et al. 2001; Pinto et al. 2006). The EST sequences are found to be more conserved among the different genotypes, thus giving rise to monomorphic banding pattern in many cases. The conserved nature of the EST sequences may be restricted to the flanking sequences with its absence in the sub-repeat units of the sequence [reviewed in Cordeiro et al. (2001)].

The comparatively limited availability of microsatellite markers in the functional domain led to the use of Sugarcane Enriched Genomic Sequences (SEGMS) and Unigene Sequences (UGMS) to develop microsatellite markers (Parida et al. 2009, 2010; Cordeiro et al. 2000) for diversity analysis. Many sequences having homology to retrotransposons could be identified in sugarcane also (Parida et al. 2009). This confirms the earlier reports of abundant distribution of retrotransposons in sugarcane (Rossi et al. 2001). Several sequences with homology to sugar gene related sequences have also been identified from these SEGMS. A correlation between the type and length of repeat motif and level of polymorphism revealed was evident, with the Class I tetra and di nucleotide repeat motifs detecting a higher level of variability than the Class I tri-nucleotide and Class II repeats. Even though the level of transferability among the cereals as a whole is low for SEGMS, transferability as high as 93.2% was observed for this class of markers among *Saccharum* species and related genera (Parida et al. 2009).

Markers from unigene sequences have also been used for germplasm studies and genome characterization in sugarcane (Parida et al. 2006, 2009). Polymorphism studies among members of cereal species have revealed many peculiar features. Among rice, wheat, maize sorghum and barley, high degree of conservation and cross transferability were evident. A set of Conserved Orthologous Set (COS) markers were identified for use across the cereal genomes which may be useful in sugarcane also (Parida et al. 2006). Functionally relevant unigenes studies in sugarcane (Parida et al. 2009) revealed that the frequency of microsatellites in sugarcane was lower than that of rice, sorghum, barley and maize. Frequency of mononucleotide repeat motifs was also less in sugarcane. About 39% of the sequences corresponded to the sugar metabolism gene sequences and the polymorphism revealed pointed towards possible variations with respect to these genes. Microsatellite repeat variations in some of the Indian clones were identified with respect to (AG)n repeats between some tropical and subtropical varieties. Among the varieties from India, the tropical sugarcane varieties contained $(AG)_{18}$ microsatellite repeats whereas the sub-tropical varieties had $(AG)_{10}$ microsatellite repeats. This variation in the number of repeats among the tropical and sub-tropical varieties may have some significance in their adaptation to the different conditions in the two zones. The fact that interspecific and intergeneric polymorphism were also revealed, makes these class of markers suited to various applications like diversity studies, hybrid identification, assessment of gene transfer to desirable genetic background etc. The use of unique gene sequences generally calls for amplification of unique bands which does not seem to be the fact in these studies, obviously due to the larger number of copies of gens/loci present. The possibility of gene silencing taking place in the multi-copy loci present, has also been speculated by the authors. Many workers have predicted from their molecular diversity analyses that the genetic base of Indian varieties is narrow, due to the limited number of genotypes that have been utilized in crossing programmes (Nair et al. 1999, 2002; Selvi et al. 2003, 2005). Quite contrary to this general belief, Parida et al. could gather evidence of a wide range of genetic similarity (0.33 to 0.84) among Indian cultivars. Thus, the genetic base of Indian sugarcane varieties may not be very narrow as predicted by earlier workers (particularly in the genic genomic regions), as it is evidenced by the unigene-based marker investigations. This is a classic example of the utilization of advanced techniques and marker systems that can reveal new information leading to novel conclusions even though the limited nature of such studies compels one to exercise caution. The transferability of these SEGMS and UGMS derived microsatellite markers to other germplasm was also tested (Liu et al. 2011). The fact that 70% of these SSR markers amplified PCR products from the genomic DNA of a Louisiana variety compared to the 30% transferability of ISMC SSR markers to US sugarcane cultivars (Pan 2006) is an indication of the role of Indian sugarcane varieties in the lineage of elite clones of other countries.

Sugar Gene Sequences for Molecular Diversity Studies

Sequences from sugar genes have also been specifically used for polymorphism studies in sugarcane. EST derived RFLP fragment cDNA for sucrose synthase could demarcate the high and low sugar clones in a population, thus pointing towards the possibility of these being used as markers for sugar content (da Silva and Bressiani 2005). Lingle and Dyre (2004) identified polymorphism in the promoter region in Sucrose Synthase 2 gene and speculated that a short simple sequences repeat in intron 14 may be polymorphic among different sugarcane genotypes. TRAP marker system as a tool for diversity studies and genome characterization (Alwala et al. 2006a, b) has also helped in identifying species and genus specific bands, thus making them useful in fingerprinting and gene introgression studies. This has involved sucrose metabolism genes as fixed primers thus revealing the diversity with respect to sugar sequences in the clones. The many bands identified with respect to the genus and/or species and/or cultivars open up opportunities for their use in varietal identification or fingerprinting and gene introgression studies in commercial breeding programmes.

Sequence variation within a gene was also studied in the complex genome of sugarcane (McIntyre et al. 2006). Molecular cloning of Sucrose-Phosphate Synthase cDNAs and comparative analysis of gene expression revealed a difference in the regulatory phosphorylation site between two genes SoSP1 and SoSp2 (Sugiharto et al. 1997). A total of ten SPS gene family III alleles were identified in a mapping population of Q 165 × IJ76-514. Such information can be effectively utilized in breeding for quality improvement. A fine analysis of 6-Phosphogluconate di-hydrogenase (Pgd) by Grivet et al. (2001) revealed sequence diversity in different sugarcane genotypes for the gene. The results suggested the existence of two Pgd genes A and B with a single SNP in A sequence and 39 SNPs in B sequence. Thus the variation among different alleles/genes with respect to the presence of SNPs within is also revealed through this study, though a preliminary reason may be the differences in the number of EST sequences available in the genes A and B. Such differences may be exhibited by sugar genes also and may be exploited in classical breeding programmes. Singh et al. (2008) also identified a set of polymorphic markers from cDNA library sequences in a bulk segregant analysis for high and low sugar segregants, suggesting their utility in differentiating high and low sugar clones.

EST based microsatellites from Soluble Acid Invertase (SAI) genes were used to identify polymorphism in this sequence among the different *Saccharum* spp.

Fig. 1 SSCP patterns obtained with the microsatellite markers SAI-MS Lanes *1–6 S. officinarum* clones, lanes *7–11 S. barberi* clones, lanes *12–16 S. sinense* clones and lanes *17–21 S. spontaneum* clones. M_1 gene ruler 50 bp DNA ladder (denatured). M_2 gene ruler 50 bp DNA ladder (non-denatured)

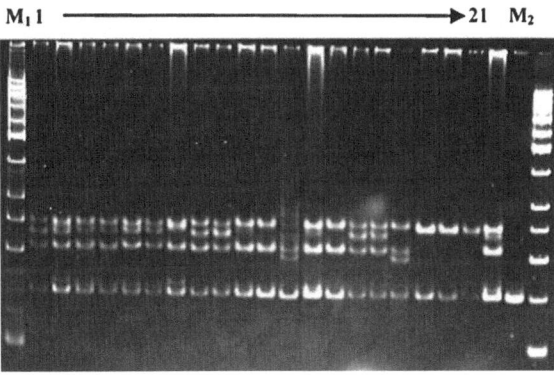

Fig. 2 Polymorphism among some high sugar clones using EST derived primers. Lanes *1–4* have the same parentage and lanes *5, 6* have the same parentage

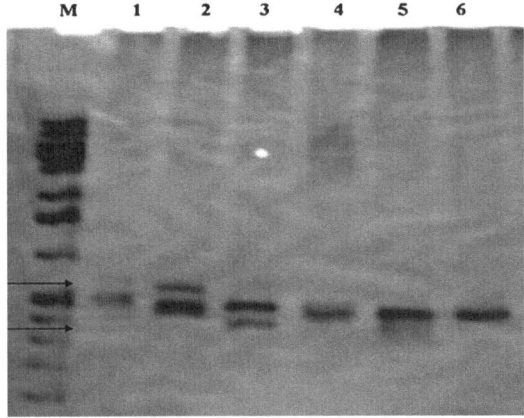

clones and species hybrids (Swapna et al. 2011a). Even though low level of variation was revealed among the Indian cultivars, the species level clones showed polymorphism esp., with respect to the *S. spontaneum* clones when Single Strand Conformation Polymorphism (SSCP) was employed (Fig. 1). SAI has been suggested to be an enzyme that has an important role in inversion of sucrose in the immature culms of sugarcane. Thus these differences assume significance in any study involving sugar variation.

Variation was revealed among some high sugar breeding stocks from India using sugar sequence specific primers (Swapna et al. 2011b). Some breeding stocks with the similar range of Pol% juice (17–19%) and same parentage in some cases, exhibited differential banding pattern (Fig. 2). There is every possibility that such variations may be due to different genes or alleles of the same gene. Such high sugar parents showing polymorphism offer excellent source for accumulation of different alleles that may contribute to increased sugar content in the progeny. This can also be utilized in pre-breeding.

Diversity Array Technology (DArT) based markers are the most advanced marker system used in the highly complex polyploid sugarcane (Heller-Uszynska et al. 2011). DArT analysis demonstrated that the markers can effectively discover and score a large number of polymorphisms. The genetic relationship revealed confirmed a strong differentiation between *S. officinarum* and *S. spontaneum*, and the ancestral species of the modern cultivars of Australia. The placement of the *S. officinarum* clones was in agreement with the idea that almost 80% of the genome of modern cultivars is derived from *S. officinarum*. Almost one-third of these markers had similarity with sugarcane EST sequences originating from the transcribed portion of the genome. Once sequencing of these markers are completed polymorphism with respect to sugar genes may also be revealed, thus enhancing the utility of these markers in sugar related studies.

5.2.2 Linkage Mapping in Sugarcane

An important strategy for unravelling the complexity of sugarcane genome will undoubtedly comprise of constructing saturated linkage maps and locating important trait-linked loci at different positions on this map. Efforts in this direction started long back, in early 1990s. Wu et al. (1992) had discussed in detail, the theoretical aspects of genetic mapping in higher polyploids. Here the pairing behaviour may be unclear in many cases. Genome duplication and other complexities add to the problems. Construction of maps directly on polyploids is difficult due to (i) large number of genotypes for each probe (ii) the higher ploidy level-auto or allopolyploidy, or a mixture in some cases (Wu et al. 1992). For some polyploid species, diploid relatives may be used as in the case of wheat (Kam-Morgan and Gill 1989) and potato (Bonierbale et al. 1988). In sugarcane the closest diploid relative is sorghum which is being utilized now-a-days for comparative genetic studies on a large scale.

In polyploids the segregation of each DNA fragment/allele can be analyzed based on its presence or absence in a particular genotype. Thus, single dose, double dose or triple dose fragments can be identified based on the segregation ratios obtained for the different markers. Of these, single dose markers (or SDRF markers) are the most useful for construction of linkage maps in polyploids. A family size of 75 plants has been suggested to be enough for detection of single dose markers and linkages in coupling phase with 98% confidence level. Also, this family size can detect repulsion phase linkages in autopolyploids with large error variance. These single dose markers can also aid in distinguishing auto and allopolyploids and for identifying preferential chromosome pairing in this crop. Suitability of different mapping population was also discussed by the authors, with a haploid population from a highly heterozygous plant being considered to be the most efficient. A hybrid population from a cross between a heterozygous parent and a haploid/homozygous parent is the next best option. Selfs of the heterozygous parent can also be used although this has been rated the least desirable. Since most of the sugar related mapping studies have used the basic information from linkage mapping carried out earlier by the different groups, the linkage mapping by various workers will also be dealt with some detail here.

The earliest attempt to construct a linkage map in sugarcane may be credited to da Silva et al. (1993) using RFLP markers on a population of anther culture derived haploid progeny of SES 208 and a cross population of ADP 068 (a doubled haploid derived from anther culture of SES 208). The map comprised of 216 loci distributed among 44 linkage groups. The estimated genome coverage was 86% with at least one marker at every 25 cM distance. This was followed by detection of single dose polymorphism using PCR based arbitrary primed reactions in the same cross population by Al Janabi et al. (1993). Linkage analysis using these SD markers resulted in a linkage map with 42 linkage groups. The predicted coverage was 85% with a marker at every at 30 cM. The finding that SES 208, (and for that matter *S. spontaneum,* for which the haploid chromosome number x = 8 had been predicted before through cytological studies), is an autopolyploid with x = 8 was derived from this study. Later these two maps were combined together to develop a consensus map (da Silva et al. 1995) using single dose, double dose and triple dose markers. The map had 64 linkage groups with a predicted genomic coverage of 93%, and an average interval of 6 cM between the markers. Thus this consensus map can be considered to be more saturated than the two previous maps. As in the other two maps here also, repulsion phase linkages could not be detected. SES 208 displayed polysomic segregation pointing towards autopolyploid nature. The auto-octaploidy was confirmed in this map also. The domesticated species of *S. officinarum* also has been mapped with a population of a cross of a representative of *S. officinarum* and its putative wild ancestor *S. robustum* suggesting incomplete polysomy (Al Janabi et al. 1994). A linkage map using RAPD single dose markers was constructed by Mudge et al. (1996) also, from the progeny of *S. officinarum* × *S. robustum.*

Attempts to utilize the cultivated genotypes for mapping studies were initiated by Grivet et al. (1996). An RFLP map using the selfed cultivar R570 was constructed using a set of 128 RFLP probes and one isozyme. 480 markers identified 96 linkage groups. A tentative *Saccharum* composite map was constructed. No significant structural differences between the genome portions inherited from *S. officinarum* and *S. spontaneum* were uncovered. A significant revelation was the possibility of the presence of clusters of almost non-recombining loci. Low recombination areas were identified in a previous mapping attempt also (D'Hont et al. 1994). This recombination heterogeneity present in the genome has great significance in the context of linkage mapping and mapping of economically important traits. Another interesting observation is the greater span of *S. spontaneum* chromosomal regions identified in these mapping studies, compared to that of other components. A possible explanation offered was that the observed level of polymorphism revealed using the above-said marker systems was more in the *S. spontaneum* genome than that in *S. officinarum* genome, due to more number of polymorphic markers derived from *S. spontaneum.* As long as more polymorphic markers are not detected from *S. officinarum* component, their utility for mapping purposes will lag behind. This again has been attributed to the 2n + n chromosomal segregation reported in the *S. officinarum* × *S. spontaneum* crosses. Garcia et al. (2006) also used commercial cross progeny for linkage mapping using maximum likelihood approach.

With the advancement in the different types of marker systems developed, the linkage maps constructed also advanced in the level of saturation and information revealed. The self population R 570 was used for genome mapping using AFLP markers (Hoarau et al. 2001). Using 37 primer combinations, 887 simplex markers were assigned to 120 co-segregation groups (CGs), with 52 unlinked markers. The cumulative length of the CG was 5,849 cm with an average distance of 6.5 cM between two markers. Even though there was considerable improvement with respect to the coverage and information revealed, this map also is not completely saturated. A tentative determination of the marker origin was also done and this was comparable with some earlier results using GISH (D'Hont et al. 1996). A very interesting observation was that, most probably, GISH efficiently identified *S.spontaneum* genome portions but its resolution power may have been insufficient to identify all recombined chromosomes. Thus the mapping information obtained with respect to recombination of genomes can be combined with the earlier information from molecular cytogenetic studies for a better understanding of the genomic constitution of cultivated sugarcane clones. Unlike many earlier maps, the AFLP based map could detect 13 cases of repulsion linkages. Some preferential pairing identified pointed towards incomplete polysomy, even suggesting the possibility of complete local disomy. The increase in the number of markers and number of progeny used helped in better coverage and higher power of detection.

Simultaneous use of AFLP and SSR markers in a segregating population of IJ 76-514 × Q 165 (Aitken et al. 2005) provided an extensive map coverage for the sugarcane cultivar Q 165. Forty AFLP and 72 SSR primers were used to generate 967 single dose and 123 double dose markers. Out of a total of 136 markers, 127 markers were grouped to eight Homologous Groups (HGs).The presence of *S. spontaneum* chromosomes homologous to two separate sets of smaller chromosomes with *S. officinarum* origin suggested the existence of ten HGs for *S. officinarum*. The bi-specific origin of sugarcane and the different basic chromosome number for the two species have been suggested by Grivet and Arruda also (Grivet and Arruda 2001). Use of additional double dose markers and 3:1 segregation markers enabled identification of an additional 18 LGs to the map. The double dose markers helped to identify the large scale duplication of chromosomes in this crop, thus bringing out the utility of these markers for polyploid linkage mapping. This study detected more repulsion linkages than the earlier results and also suggested translocations. Also a partial preferential pairing was speculated within an HG with a possibility of recombination. From a comparison of the map generated on R570 using EST-derived RGAs (Rossi et al. 2003), the authors suggested that genetic maps produced for different cultivars may reveal different chromosome arrangements even when some common markers are used. This may be due to translocations and other chromosomal remodelling including recombinations between HGs (Aitken et al. 2006). Such observations will have important implications in molecular marker application studies.

A similar attempt to develop a genetic linkage map for IJ76-514 using both simplex and duplex markers was made in a segregating population of Q 165 × IJ 76-514 (Aitken et al. 2007). The same set of primers was screened revealing 595 markers. Out of these only 240 (40%) were simplex markers showing (1:1) segregation, again demonstrating the availability of less number of single dose markers for *S. officinarum*. One seventy eight of these were distributed in 47 LGs with 62 unlinked markers. An additional 234 duplex markers and 80 bi-parental simplex markers generated a total of 123 LGs. Using the multi-allele SSR markers, repulsion phase linkages and alignment with Q165 linkage maps were identified with ten HGs from 105 LGs. Repulsion phase linkage analysis indicated that IJ 76-514 is neither a complete polyploid nor an allopolyploid. Also, occurrence of limited preferential chromosome pairing was suggested in this species. An approximate coverage of 60% was observed. Use of EST markers for linkage mapping in a commercial cross SP80-180 × SP80-4966 was attempted by Oliveira et al. (2007). 72.5% single dose markers were generated 192 Co-segregation groups (CGs) were identified from which, 120 could be grouped into 14 homology groups (HGs). Putative functions were assigned to 113 EST-SSR markers and six EST-RFLP markers based on BLAST studies. Thus this map is populated with functionally associated markers. Alwala et al. (2008) used TRAP, SRAP and AFLP markers for genome analysis and linkage mapping in the progeny of an interspecific cross in sugarcane. The framework map comprised of 146 linked markers in 49 linkage groups for *S. officinarum* and 121 linked markers in 45 linkage groups for *S. spontaneum*.

Of late, DArT marker systems have been used in molecular analysis in a population from Q165 × IJ 76-514 (Heller-Uszyneska et al. 2011). Even though a complete linkage map using DArT markers is still to be reported, the polymorphic DArT markers were incorporated into the linkage map generated previously (Aitken et al. 2005). Majority of the markers mapped to similar regions of the genome as in AFLP and SSR marker generated maps.

5.3 Mapping of Sugar Related Traits in Sugarcane

As we all know, sugar metabolism involves a large number of players. Other components like stalk girth, stalk height, number of millable canes etc. are also important in the final sugar yield. Thus there may be a large number of genes/loci which influence the trait. These may be spread on different homo(eo)logous chromosomal regions in the entire genome. Hence multiple doses of each loci may be present. Some of these may be linked to single dose markers, while some would necessitate the application of double (or higher) dose markers. Thus a simultaneous utilization of different types of markers will help in a better coverage. Another important aspect is the temporal regulations in the sugar accumulation and sugar content. As the crop begins to ripen one can identify early varieties, that give a reasonably high sugar content quite early in the season. Such early varieties, if cultivated fetch an additional incentive to the farmer. Then there are the mid-late varieties that meet the above-said criteria later during the season. Thus there are temporal regulatory factors at play in sugar accumulation and cane ripening.

Different loci may have roles to play here. All these factors need to be taken care of in studies involving mapping for sugar related traits.

Among the trait specific mapping studies carried out so far, mapping for sugar related traits have been exploited to the maximum by many groups. Other reports were on association of DNA markers with disease resistance (Daugrois et al. 1996; Asnagahi et al. 2004; Al-Janabi et al. 2007) and flowering time (Guimaraes et al. 1997). The earliest attempts to map sugar related traits in sugarcane was carried out by Ming et al. (2001, 2002) for sugar content and sugar yield related traits.

Two interspecific segregating population (i) a set of 264 plants from *S. officinarum* Green German (GG) × *S. spontaneum* IND-81-146 (IND) (ii) a set of 239 plants from *S. spontaneum* PIN 84-1 (PIN) × *S. officinarum* Muntok Java (MJ) were used for sugar related mapping studies by Ming and his group. A total of 36 marker trait associations were identified—14 from GG × IND (8 from GG, 6 from IND) and 22 from PIN × MJ (18 from MJ, 4 from PIN). The eight GG QTLs explained 38.6% of phenotypic variation for sugar and six IND QTLs explained 36% of variation. A total of 65% of the variation was accounted for by the 14 QTLs. Three QTLs explained a portion of transgressive variation observed. The 22 QTLs from PIN × MJ contributed to 68.3% of total variation with 18 MJ QTLs alone contributing for 45.7% of variation and the four PIN QTLs accounting for 33.4% variation.

Comparison with the sorghum linkage map revealed that the 36 sugar QTLs corresponded to eight non-overlapping regions of the sorghum genome. In 75% of these regions both high sugar and low sugar QTLs were detected, thus confirming the importance of such loci in breeding for sugar content. The appropriate utilization of these QTLs identified will depend on a proper analysis by which QTLs with exceptional effects—i.e., QTLs in high sugar parent which decrease sugar content and those from low sugar parents which increase sugar content- can be detected. Dosage studies led to the conclusion that even though more than one loci or more than one allele may be present, multiple doses of loci with favourable alleles do not always lead to increased phenotypic expression, thus suggesting a non-linear interaction or epistasis among the loci (Ming et al. 2001; Eshed and Zamir 1996). As suggested by the authors, such interactions may exist among non-linked loci also. It may be that a single copy of these multiple doses itself is sufficient enough to give a desirable level of phenotypic expression. Thus incorporation of a single copy of multiple alleles for increased phenotypic expression is a possibility in this polyploid. An interesting hypothesis proposed by the authors is the evolutionary significance of these non-additive multi-dose QTLs and the possibility of their role in imparting stability across environments. Among the 36 QTLs, candidate genes were detected only in 17.5% cases. The extent of association between candidate genes and QTLs varied among the sources (strong for GG QTLs and non-existent for PIN QTLs with those from IND and MJ in between). The same population was used to map QTLs for other sugar related traits by Ming et al. (2002). QTLs for sugar yield, pol, stalk weight, stalk number, fibre content and ash content were searched with 735 DNA markers. One hundred and two significant associations were mapped with 61 linked to QTLs and 41 unlinked.

Fifty out these 61 mapped QTLs fell into 12 LGs of seven sugarcane HGs. These 50 QTLs have been grouped by the authors into different categories depending on the regions where they were mapped. Similar conclusions were arrived at, regarding transgressive segregants, dosage effects etc. as that was in the earlier mapping study.

The selfed population of R570 was used for mapping sugar QTLs by Hoarau et al. in 2002. Using 1,000 AFLP markers 45 Quantitative Trait Alleles (QTAs) were mapped linking to four traits-stalk length (SL), Stalk diameter (SD), number of millable stalks (NS) and Brix (BR). Single marker QTAs (S-QTAs), multi-marker M-QTAs) and linear interactions between simplex markers (int-QTAs) were analyzed in this study. Out of these 11 S-QTAs were detected for brix, 10 for SD, 13 for SL and 11 for NS. Two S-QTAs were common in both the 2 years studied for BR whereas one each was common for the other three traits, thus making a total of five S-QTAs identified across the years. The digenic interactions detected helped to illustrate epistasis which otherwise is difficult to assess in sugarcane. The consistency across different years was low for the QTAs. This was the first sugar QTL study in a cultivated sugarcane variety. In another QTL analysis using the population from the cultivars Q 117 × MQ77-340 (Reffay et al. 2005), a total of 23 Marker-Trait Associations (MTAs) were identified for CCS, Pol and Brix, using ancestor-identified/known ancestor-origin markers. Only one was found to be associated with all the three traits over the years. The study also helped identify specific genomic regions contributed by specific ancestor (i.e., Mandalay) to elite parental clones.

QTL affecting sucrose content in the Australian cultivar Q 165 were studied by Aitken et al. (2006). The same population and set of markers were used as that for linkage mapping (Aitken et al. 2005). Brix, Pol and commercial cane sugar (CCS) were included as sugar traits. Both single dose and multi dose markers were used for QTL detection. A total of 37 QTLs for brix and pol were identified, of which eight were specific to early period and nine specific to mature cane. Each QTL explained 3–9% of the phenotypic variation for the two traits. Thirty out of the 37 QTLs were mapped into 12 genomic regions, which in turn were spread over six HGs. Thus 12 loci related to sugar related traits were mapped in this population. Repeatability among the 2 years of study was not evident even though six of the QTLs were involved in interaction in both the years. Also consistency with respect to the direction of the effects was evident. The same population was used for yield related trait mapping by the group (Aitken et al. 2008). The mapping was done for stalk number (SN), stalk length (SL), stalk diameter (SD) and stalk weight. Both simplex and multiplex markers were used Composite Interval Mapping (CIM) was also carried out in this study. The markers could be located to 27 genomic regions on 22 linkage groups and six HGs. Except in two cases, a group of significant markers on one LG was considered as one Quantitative Trait Allele (QTA). Thus there was a total of 22 QTAs with these individually accounting for 4–10% of the phenotypic variation. A number of QTAs (1–3) at each QTL contributed to a particular trait. These QTAs grouped into 16 genomic regions or QTLs, mapped by SSRs. The variance explained in total ranged from 15 to 36%, with the value going up to 60% with the inclusion of digenic interactions. Evidence for epistasis was strong in this study also as in many previous ones. As in earlier studies (Ming et al. 2002) association of a QTA with more than one trait was detected here also but the level of significance varied in different interactions. Comparison with a previous mapping attempt by Aitken et al. (2006) for sugar related traits revealed that nine stalk-related QTAs mapped to similar locations to that of the twelve mapped sugar related QTLs in the same population (Ming et al. 2002; Hoarau et al. 2002). Even then, very poor correlation was detected among stalk traits and sugar related traits. The lack of correlation exhibited between sucrose content and yield components in this study prompts one to rethink the general notion among the breeders that increase in sucrose content mostly leads to a decrease in yield (even though more studies may be needed to confirm this). An analysis of the positive and negative effect QTAs indicated that in this population, an average of four QTAs with positive effects and 1.6 QTAs with negative effects combined together to give rise to individuals having the highest values for stalk weight. Thus it is possible to select for desirable traits by maximising the positive QTAs with additive effects, while minimizing those with negative effects.

Sugar related traits (Brix and Pol) were mapped using AFLP, SRAP and TRAP markers using the progeny of a cross *S. officinarum* Lousiana Striped × *S. spontaneum* SES 147B (Alwala et al. 2009). A total of 41 putative QTLs (30 from *S. officinarum* and 11 from *S. spontaneum*) were identified for the two traits. Individual QTLs explained a phenotypic variation of 15.1–21.6%. Nine digenic interactions (iQTLs) were also detected. Seven QTLs were consistent across the years of study and two QTLs were unique to early plant growing season. The utility of discriminant analysis (DA) to detect marker-trait association in this polyploid species was also tested by the authors. Some markers either identical to or localized to the vicinity of QTLs were identified by DA. Some unlinked markers also could be detected by this method, there by suggesting the use of DA also as a complementary method of mapping in this crop. Pinto et al. in his QTL mapping attempts using RFLP-ESTs could associate a sucrose synthase derived marker with a putative QTL having a high negative effect on cane yield and also with a QTL having a positive effect on Pol in two crop cycles (Pinto et al. 2010). Thus different marker systems have helped in locating a number of sugar related QTLs in different sugarcane populations.

5.4 Mapping and Marker Assisted Breeding (MAB): Need for a Cautious Approach?

Genetic recombination forms the basic principle of genetic linkage analysis which in turn is used for detection of genes and QTLs (Tanksley 1993). Appropriate statistical methods can be applied to detect trait linked DNA markers for a specific trait and chromosomal region or loci can be located on a linkage map (Kearsay 1998; Paterson et al. 1988). Once DNA markers linked to QTLs are identified and mapped to different genomic regions, these QTLs are to be validated. Thus preliminary QTL data may not be always applicable as such to the practical field situations for marker assisted breeding. At the initial stages of QTL mapping studies, researchers were highly optimistic about the scope of marker assisted breeding in different crops. But with the progress of time and advancement in knowledge in this field, it appears that results of Marker Assisted Selection (MAS) and related strategies are not very much visible in the actual plant breeding programmes (reviewed in Collard et al. 2005, 2008; Hospital 2009). This is not specific to sugarcane even though the situation is slightly different in other less complex crops like rice. With respect to qualitative traits there are some examples where QTL identification and subsequent steps have been actually applied in field situations as in the case of soybean cyst nematode screening and Quality Protein Maize (QPM).

Variation exists as to the ease with which linkage mapping and QTL mapping can be accomplished in different crops. Genome complexity of the crop in question is a major point of consideration. As is well known, the highly heterozygous complex polyploid nature of sugarcane has been an important deterrent for such mapping studies till early 1990s. The near absence of a close diploid relative as in many other cases delayed the successful development of a well saturated linkage map. With the designation of *Sorghum* as the closest relative and with advancement in molecular mapping techniques, sugarcane too, has been included in the elite group of crop plants whose linkage map is gradually getting saturated, there by leading to mapping of several trait linked markers also.

Having studied the development of linkage maps and QTL mapping for sugar traits in sugarcane, let us have an idea about the road blocks commonly encountered in this type of studies and in sugarcane, in particular. In sugarcane the exact ploidy level of the crop has been a matter of debate for a long time. Of late some of the mapping studies have confirmed the ploidy number of $x = 10$ for *S. officinarum* and $x = 8$ for *S. spontaneum* species. With such a high ploidy level, development of haploids for mapping studies is not a very feasible option. Sugarcane is not very responsive to anther culture techniques and other haploid generation strategies. Thus the use of haploids for construction of genetic maps is not an easy alternative. Till now SES 508 is the only haploid developed to our knowledge that has been used in such type of studies.

Unlike in other diploid crops where two copies of a particular locus may be detected, in sugarcane up to 11–12 alleles of a locus might get detected. These different alleles may themselves exhibit interactions among themselves as evidenced in the case of some multidose QTLs. Also, since there may be a large number of loci affecting complex traits like sugar content and yield, individual effects of these QTLs may be very little, sometimes making it difficult for detection of such minor effect loci. The absence of segregating large QTL effects for such traits may pose problems in their actual utilization, as they may result in small genetic gains.

Sugar related traits being physiologically affected to a large extent, the loci influencing such traits will be highly variable due to their interaction with the genetic background, physiological conditions etc. Thus QTL mapping for such traits will give best results if they are carried out in such backgrounds similar to that encountered in breeding programmes. Development of mapping population using cultivated clones is one approach. A possible drawback in this case may be the comparatively narrow differences among the available cultivated clones with respect to some traits like sugar content. Among the cultivated clones, all the individuals will have a minimum level of sucrose content below which it will not be cultivated at a commercial level. Thus a parallel programme for identification of low sugar parents and their utilization in parental line development will ensure a population with a fairly wide range for sugar content. Needless to say, the already reported mapping studies have a reasonable range with respect to the particular phenotype and have served the purpose exceedingly well.

Compared to the other diploid crops where linkage maps can be developed for both the parental lines, in sugarcane the maps have to be constructed for the two parents separately. The genomes of cultivated clones have a constitution of approximately 80% *S. officinarum* genome and approximately 10% of *S. sponta-neum* genome (D'Hont et al. 1996), with almost 10% constituted by recombinants. The mapping studies point towards a difference with respect to the two components. It has been observed by many workers that the extent of coverage of *S. officinarum* genome in the map is low compared to the *S. spontaneum* component. This has been attributed to the low level of polymorphic single dose markers identified in this part of the genome. This observation calls our attention to an important cytological phenomenon that occurs in *S. officinarum* × *S. spontaneum* crosses (Bremer 1961) or even in *S.officinarum* × a commercial clone (Piperidis et al. 2010). The 2n + n transmission or the female restitution taking place in such situations may be one reason for the low level of polymorphism in the *S.officinarum* component of the genome. Thus, the availability of lower number of single dose markers for mapping purpose limits the extent of coverage in this part of the genome.

Another peculiarity in the plant genomes, is the existence of genomic regions with variable levels of recombination. Such recombinational hot spots (with increased recombination frequency) or cold spots (with lower frequency of recombination) exists in sugarcane genome also (D'Hont et al. 1994; Grivet et al. 1996) as reported in other plants. This in turn, may determine the uniformity with which the existing trait-linked loci can be located. Temporal variations may also exist where the once identified hot spot may later become a cold spot or vice versa. This may cause alterations in the population developed there by leading to disappearance of some associations or appearance of new associations. This will also determine how effectively the mapped loci can be used in breeding pro-grammes. In sugarcane cytological peculiarities have been reported like en-masse chromosomal elimination or segmental eliminations [reviewed in Sreenivasan et al. (1987)].These may also act as a cause for genomic instabilities leading to complications in marker assisted breeding applications. These may also result in emergence new marker-trait associations or disappearance of some linkages in future. Rossi et al. (2001) had identified 21 differentially expressing transposable elements (TE) in sugarcane from the EST sequences developed in SUCEST project, which are potentially active. A search for TEs revealed the presence of 276 clones out of 2,60,781 sequences (81,223 Phraps) that had homology to previously reported TEs. There is a remote possibility that some re-arrangements in small portions in the genome or in the whole genome may take place in the course of time, due to the activity of these TEs. Such phenomenon can also be a cause of genomic instability giving rise to some discrepencies in QTL mapping results, in the long run.

Effect of genetic background is an important factor determining the success of the QTL in MAS and other applications (or for that matter its effectiveness in a different genetic background). This is increasingly true in the case of traits like sugar content as mentioned before. Such situations have been encountered in other crops like rice and tomato (Liao et al. 2001; Lecomte et al. 2004; Chaib et al. 2006; Causse et al. 2007). Sometimes this may be due to the small individual effects of the QTLs identified. This may be particularly applicable to sugar traits in sugarcane, where large number of genes, each with a minor effect, determine the final outcome. Such loci when transferred in a different background may be difficult to detect as observed by Charcosset and Moreau (2004). Also due to the selection for high sugar and fixation of the loci, mapping studies have shown that segregating larger chromosomal regions are very rare in sugarcane thus posing problems as they may result in small magnitude of genetic gain. Interactions among QTLs is also another factor that may influence the ultimate success of mapping programmes. In many sugar related multi-dose QTLs non-linear allelic interactions have been observed. These alleles when used in introgression studies may not give the expected level of expression even if all copies are introgressed. Thus the optimal copy number may be arrived at before application of QTLs for gene pyramiding. Also the QTLs, both with normal and exceptional effects, may demand a detailed analysis before these are put to use for gene introgression studies.

Even though the initial assumption was that QTLs are least affected by environment, recent evidences suggest that environmental effect do have a role to play in the success of MAS using QTLs. The magnitude and direction of QTL effects may vary in different environment. Such results are reported for sugar related traits in sugarcane also where across-the year-stability is not evident in some cases. Approximately 40% of the sugar content QTLs could not be detected in the same population over different years (Aitken et al. 2006) and in some yield related traits it depended on the significance level employed (Aitken et al. 2008) There were differences in the level of detection at a genome wide level of 5% and at an individual detection level of 5%, pointing towards a high genotype × crop cycle interactions for yield related traits. As mentioned before, both early sugar accumulating and late sugar accumulating varieties exists in sugarcane, with the former fetching more monetary gains for farmers and also ensuring an early and longer crushing season for the factories. These temporal differences are reflected at QTL level also. No QTLs linked to both the early and late sugar accumulation could be detected in one such study (Aitken et al. 2006). This adds a new dimension to the sugar related QTLs mapping studies. This also needs to be kept in mind while planning a MAS strategy for sugar accumulation/sugar content in this crop.

With the current level of technological advancements it has become possible to overcome many of the problems in sugarcane. But still some difficulties pertaining to the complexities of the crop will have to be encountered as we go along our programmes. One has to also consider the traits that are suitable to be mapped, the population to be used for mapping, the types of markers to be used in the study etc., so that the results can be used effectively in marker assisted breeding programmes.

5.5 Comparative Genetics and Genomics

Comparative genetic studies form an important strategy that can be effectively put to use in the case of a highly polyploid, complex crop in sugarcane. In several crops the concept of utilizing the information from a much simpler crop and applying it to the target crop of interest which is more complex, had caught the attention of researchers. In sugarcane not much advance was made in this area till the 1990s. Gradually, with the concept of "grasses as a single family" taking root, comparative studies were initiated with other crops of grass family like rice, maize, sorghum etc. The finding that sorghum may be more similar to sugarcane than other crops led to the demarcation of sorghum, also a C_4 plant, as the closest diploid relative of sugarcane. Since then several studies have been carried out in the area of molecular diversity, cross-transferability of molecular markers, comparative genome mapping, comparative functional genomics etc., utilizing the information generated from sorghum.

Within the Andropogonaea tribe, early reports focussed on crops like maize and sorghum (Hulbert et al. 1990; Whitkus et al. 1992) demonstrating that many orthologous loci were common to the two. In sugarcane the first reports of comparative genetic studies were with maize (D'Hont et al. 1994). The results indicated that comparative analysis using simpler and low ploidy level crops like maize may help in unravelling of the complexities of sugarcane. Sugarcane, maize and sorghum genomes were compared in some studies (Grivet et al. 1994; Dufour et al. 1996, 1997; Glaszmann et al. 1997; Guimaraes et al. 1997; Ming et al. 1998). These studies confirmed that sugarcane and sorghum are more closely related to each other than either of them is to maize. Comparative studies by D'Hont et al. (1993) demonstrated synteny between maize and *Saccharum* for several chromosomal segments. Perfect collinearity at several locations between the genomes of sorghum and *Saccharum* was reported by several workers (Grivet et al. 1994; Dufour et al. 1996; Guimaraes et al. 1997). Large chromosomal rearrangements like translocations, centric fusion etc., could also be detected in some cases and these were presumed to have played a significant evolutionary role. Conservation of physical structure of genome fraction was also suggested. Such similarities provide some insight into the organization of a common ancestral genome from which duplications, divergence and other modifications might have taken place. A lower rate of chromosomal recombination was also speculated for sugarcane in these studies (D'Hont et al. 1994; Grivet et al. 1994). A simple pictoral

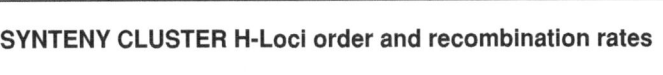

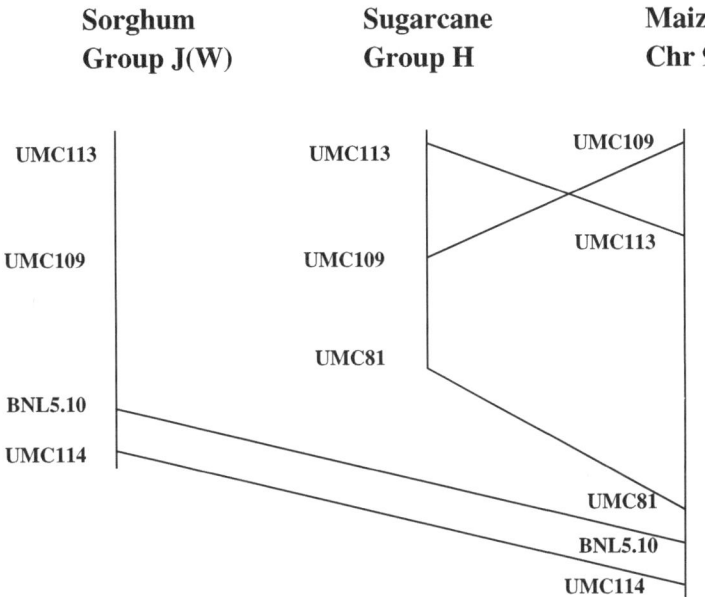

Fig. 3 A pictorial representation of the comparative analysis of genomic regions in sorghum, sugarcane and maize in synteny cluster H (Adapted from Grivet et al. 1994)

representation of one such comparative study is given in Fig. 3 showing the synteny and collinearity with respect to sugarcane and sorghum genomic regions. In maize there are rearrangements like inversions with respect to some loci.

Cross-transferability studies with different molecular markers also helped in such comparative studies. The fact that several of the markers tried in one crop amplified in other crops and could be located in both the genomes was an indication of the similarities existing at many loci among these crops. Selvi et al. (2003) demonstrated the transferability of maize microsatellite markers to sugarcane and related genera. The increased transferability of EST-SSR derived markers compared to genomic SSR derived ones across *Saccharum* species and related genera was confirmed by many workers (Cordeiro et al. 2000; Pinto et al. 2006). These also led to experimental confirmation that many of the expressed portions of the genome are highly conserved across the Poaceae family. Unigene sequences also revealed synteny of expressed chromosomal regions in different members of the grass family like sorghum and maize (Parida et al. 2006). Such conservation among the different species level clones was also exhibited by microsatellite markers from Soluble Acid Invertase EST sequences (Swapna et al. 2011a).

Table 1 Comparative mapping of sugar QTLs (including putative ones) from sugarcane (Ming et al. 2001)

Sorghum linkage group	QTLs/candidate genes mapped to the linkage group		QTLs with correspondence to specific regions and not to candidate genes	
	Sugarcane	Maize	Sugarcane	Maize
A	5	2	4	
B	4	2	2	–
C	1	1	2	
D	2	1	2	
F	–	–	3	–
G	2	1	3	
H	1	1		
I	3	1		
J	–	–	2	2

Unigene derived Sugarcane Microsatellites (UGMS) and Sugarcane Enriched Genomic Microsatellites (SEGMS) were used (Singh et al. 2011) to study the cross transferability in nineteen accessions (including *Saccharum* spp. clones, related genera, rice, wheat, maize and sorghum). 74.07% of the primers amplified in sorghum and 92.59% amplified in maize. This confirms the hypothesis that some level of conservation exists among these genomes with respect to the functional portions as is also evident from CISP marker studies by Khan et al. (2011).

Such comparative analyses have been attempted in mapping studies also by different researchers. In an attempt to map QTLs linked to sugar content in sugarcane Ming et al., had hypothesized that candidate genes for carbohydrate metabolism in biomass crop might be identified based on mutations effecting seed development in other related crops (Ming et al. 2001). The group had earlier identified 84% homologous loci that mapped to both the genomes (Ming et al. 1998), with only one interchromosomal and two intra chromosomal rearrangements between *Saccharum* and *Sorghum*. The comparative analysis of the sugar QTLs identified in sugarcane with respect to the different sorghum linkage groups identified by these authors (Ming et al. 2001) may be summarized as follows (Table 1)

These corresponded to eight non-overlapping regions of the sorghum genome. In six of these eight regions, QTLs from both the parents (i.e., QTLs with both increasing and decreasing effects) were found to co-exist. The study concluded that seed and bio-mass crops may share common mechanisms for carbohydrate metabolism, to some extent with a partly-overlapping basis for variation. More recent QTL mapping studies for sugar content in sorghum (Shiringani et al. 2010) may aid in a better comparison of this trait in the two genomes.

The first enzyme involved in C_4 photosynthesis pathway is the C_4 phosphoenolpyruvate carboxylase (PEPC). The diversity of this enzyme was analyzed in different sub-families including Andropogoneae (Besnard et al. 2002). *Saccharum officinarum*, *S. spontaneum* and *Sorghum bicolour* were included in the study. The amplicons obtained from specific primers showed 78–99% sequence homology with known grass C_4 PEPCs. Out of the four indels identified two were specific to the tribe Andropogoneae. Further studies in this direction may help us in sugar metabolism studies in this crop.

The advanced techniques of BAC library construction, screening and sequencing has helped in successful application of comparative studies of sugarcane and sorghum genomes at a very basic level (Wang et al. 2010). The study demonstrated that sorghum genome can serve as an excellent template for the assembly of sugarcane anchor sequences, even when a high density sugarcane map (approx. two markers per Mbp) or a physical map with an excellent coverage are not available. Twenty sugarcane Bacterial Artificial Chromosomes (BACs) corresponding to each of the sorghum chromosome arms were sequenced. The genic regions of these BACs had an average sequence identity of 95.2% with sorghum. About 51.3% of these BAC sequences could be aligned with sorghum sequences collinearly. The non-aligned regions were composed of non-coding and repetitive sequences. One BAC sequence 1072L01 aligned with multiple sorghum chromosomes indicating large scale chromosomal rearrangements. Within BACs also minor rearrangements were reported between the two genomes. Some discrepancies between direct sequencing comparisons and genome estimates exists, with several possible reasons being suggested by the authors. More genes were found in the sugarcane sequence fragments than in sorghum fragment (155 from sugarcane ESTs, 28 from sorghum ESTs and 26 from sorghum annotated CDs). Thus the authors demonstrated the usefulness and strength of comparative mapping strategy in a complex crop like sugarcane were neither a high density map nor a well saturated physical map is available.

Many sequences in the sugarcane BACs were aligned with the sorghum chromosome arms, thus giving an idea about their probable locations in the genome. Putative functions for these sequences were also deduced from these investigations. Among the many sequences that were aligned in the two genomes, organization of one homeologous region of the two crops deserves special mention (Fig. 4). Sorghum Chromosome 6 was aligned with sugar BAC 204D12 contig 5. Here, among the three genes located, gene 1 has been reported to be having the maximum hits with a sugar transporter protein. An efficient transport system involving sugar transporter genes, among many others is a prerequisite for a normal sucrose accumulation process in the plant. Thus these transporters may have an important role in sugar metabolic pathway. Involvement of sugar transporters in sugar signalling mechanisms regulating photosynthesis in higher plants is well known form earlier studies also (Williams et al. 2000).

The varied level of conservation exhibited by some micro RNA (miRNA) sequences from sorghum, sugarcane, wheat and switchgrass (Zanca et al. 2010) has opened up another area of research in comparative studies. The authors could identify distinct level of miRNA expression in leaf sheath, leaf roll, lateral bud etc., with every possibility that these may be expressed in stem tissues also. These may also be involved in sugar metabolism gene expression.The relatively low level of variation in miRNA accumulation between *S. officinarum* and *S. spontaneum* may reflect genetic buffering to maintain gene expression levels. Most of the sugarcane miRNAs studied had homologs in rice and sorghum.

(a)

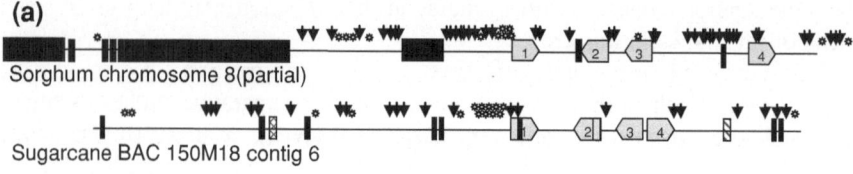

(b)

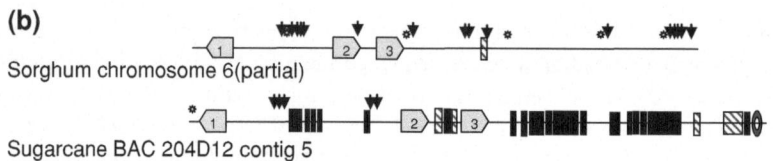

Fig. 4 Comparison between partial sequences of Sorghum chromosomes and sugarcane BACs. **a** Sorghum chromosome 8 partial sequence and sugarcane BAC 150M18 contig 6. Genes 1–4 showed maximum BLAST hits to SEU3A protein, thioredoxin M-type chloroplast precursor, chloroplast protein and exonuclease family protein respectively. **b** Sorghum chromosome 6 partial sequence and sugarcane BAC 204D12 contig 5 (showing expanded sugarcane sequences. Genes (*blue boxes*) 1–3 showed the maximum BLAST hit to sugar transporter protein, hypothetical protein OJ1065_B06.22 and expressed protein respectively. (Reproduced from Wang et al. 2010)

Thus the scope of comparative genomics has widened, from the mere comparison of genome organization or to study synteny and collinearity, to comparison of trait-specific genes, their location, their role in phenotypic variation etc. The differences that are exhibited by sorghum and sugarcane compels us to practice some level of caution in blindly fitting the experimental evidences from sorghum, to sugarcane. The basic differences in ploidy levels, gene copy number etc., will also have to be taken into consideration before applying the information generated from sorghum or other members of the grass family to sugarcane.

5.6 Functional Genomics in Sugarcane: Unravelling the Many Secrets of the Genome

In any plant, growth and development is characterized by a large number of biochemical and physiological alterations that take place during the entire growth period. In sugarcane, sucrose accumulation in the developing internodes form an important criteria that characterizes growth and development. Besides, sugarcane being a long term crop is subjected to a gamut of environmental stresses during its entire growth period which will also determine the final products in the various metabolic pathway (sugar being important among these). The recent development in the functional genomics arena undoubtfully, aid in understanding these complex regulatory mechanisms through expression analysis and other strategies

(Manners and Casu 2011; Souza et al. 2011). The information generated from functional genomics can be useful for such applications like molecular marker development. The International Rice Genome Sequencing Programme was among the first such attempts at a global level in unravelling the genetic information encoded in the plant genome. In sugarcane also such attempts at an International level has yielded very good results. The Sugarcane Expressed Sequence Tag (SUCEST) Programme involving Brazil, USA, South Africa and Australia which was initiated in 1998 was a crucial step in this direction. The SUCEST database is the largest for any monocot species (Vettore et al. 2001). The sugarcane EST project SUCEST has a database containing 2,38,000 ESTs from 26 cDNA libraries constructed from several organs and tissues sampled at different intervals (http://sucest.lad.ic.unicamp.br/en/). Studies on development of markers, RGAs and mapping for important economic traits have made use of this valuable EST database on a large scale. The Sugarcane Genome Sequencing Initiative (SUGESI) Consortium is another collaborative attempt by Australia Brazil, France, South Africa and United States, for unravelling the complexities of sugarcane genome/transcriptome, mainly through BAC sequencing strategy. The cultivar with the best characterized genome, i.e., R570 was chosen for this initiative. Recent developments in this area have led to a vast increase in knowledge of sugar metabolism and sugar accumulation in this crop.

One among the earliest strategies found to be effective was to combine cDNA subtractive hybridization combined with microarray screening for candidate gene analysis and identification (Carson et al. 2002). Among the many ESTs generated from different organs those from the internodal tissues were expected to contain information with respect to genes for sugar metabolism. Several ESTs with putative identity showed homology to cell wall metabolism, carbohydrate metabolism, stress response and regulation. About 10% showed homology to carbohydrate metabolism, but none was directly related to sugar metabolism per se. Later attempts (Casu et al. 2003, 2004) resulted in identification of novel sugar transporter homologues and other differentially expressed transcripts from maturing stem vascular tissues. Expression of trehalose-6-phosphate synthase (TPS) and trehalose-6-phosphate phosphatase (TPP) were detected. Even though the exact implication of this finding was difficult to be deduced at that time, later studies suggested a probable role for changes in TPS:TPP ratio in sugar partitioning and photosynthetic regulation in sugarcane leaves (McCormick et al. 2009). A very strong correlation of trehalose with sucrose accumulation has been detected in metabolite abundance studies in the sugarcane stem, where correlation of several metabolites with development was investigated (Glassop et al. 2007).

Anatomical and physiological modifications to maximize CO_2 fixation have been adapted by C_4 plants like sugarcane. Three probable primary carbon cycle pathways for C_4 photosynthesis have been described. These differ with respect to two aspects (i) 4-carbon organic acid intermediate transported from mesophyll to bundle sheath cells, i.e., malate or aspartate (ii) 3-carbon acid returned to mesophyll cells as well as the decarboxylation enzyme present in bundle sheath cells. This can be NADP+-malic enzyme (NADP-ME), NAD+-malic enzyme

(NAD-ME) or Phosphoenol pyruvate-carboxykinase (PEPCK) (Taiz and Zeiger 1998). Sugarcane has been classified to function using NADP-ME pathway (Bowyer and Leegood 1997). But the frequencies of PEPCK related tags compared to that of NADP-ME in a Serial Analysis Gene Expression (SAGE) study in sugarcane leaves (Calsa and Figuera 2007) led to speculations that PEPCK mediated decarboxylation pathway may predominate over NADP-ME pathway in mature sugarcane leaves though both may occur in this crop. Gene expression with SAGE could be correlated with arbitrary expression values from RT-PCR. Some level of physiological plasticity with respect to these pathways has been predicted for maize (Taiz and Zeiger 1998), which may be applicable in sugarcane also.

Potential anti-sense tags identified mostly corresponded to photosystem and chloroplast components. The decarboxylating NADP dependant malate dehydro-genase (NADP-MDH) antisense tag was more enriched suggesting a supposed association with the putative downplay of NAD-ME C_4 pathway. Lesser suppression of non-decarboxylating MDH also suggested the occurrence of NADP-ME pathway at a lower level. Regulation of metabolic activities by mRNA alternate splicing or alternate termination was also indicated through these experiments. There is every possibility that members of the sugar metabolism gene family may also exhibit such transcriptional and post-transcriptional regulatory mechanisms.

Source-sink relationship has been demonstrated in sugarcane through regulation of leaf photosynthetic activity by carbon demand from sink tissues. Leaf sugar accumulation in C_4 sugarcane may alter the expression of leaf photosynthesis genes. This was confirmed through an Sugarcane Genome Array analysis (McCormick et al. 2008a, b). This is a very powerful tool for profiling quantitative changes in gene expression in response to developmental and environmental changes. By artificially inducing sucrose and hexose accumulation in sugarcane leaves through cold girdling, these workers were able to demonstrate that 21 photosynthesis related genes associated with Photosystem I, Photosystem II, C_4 photosynthetic pathway and PCR cycle were down regulated. The regulation of TPS and TPP indicated an important role for Trehalose 6 Phosphate (T6P) in sugar signalling mechanisms. An increased expression of genes for cell-wall synthesis, assimilate partitioning, phosphate metabolism and stress were also exhibited. Pi translocators and PPases may be potential targets for manipulating sugar accumulation in sugarcane culms. The possibility that hexose rather than sucrose may be the key regulatory component in the photosynthetic activity was also suggested in the context of source-sink relationship in sugar pathway regulation. It has been reported that an increase in the hexose-phosphate concentration arising from a restriction in the conversion of hexose phosphates to triose phosphates may increase sucrose synthesis in young internodes (van der Merwe et al. 2010). A definitive role of pyrophosphate: fructose-6-phosphate 1-transferase (PFP) in sugarcane in carbohydrate partitioning has also been confirmed by these studies. The possibility of increasing sink sucrose through effective manipulation of source-sink communications has been speculated and discussed in the context of Indian cultivars also (Chandra et al. 2011), where differences exists between the subtropical and tropical regions of the country.

Table 2 Differential expressions of genes in segregating population for sugar content

Category	Level of expression		Number of genes
	High sugar clones	Low sugar clones	
Hormone biosynthesis	↑		1
Receptors	↑		1
Transcription	↑		1
No matches	↑		2
Adapters		↑	3
Inositol		↑	2
Protein kinases		↑	1
Putative protein		↑	1
Stress		↑	8
Transcription		↑	2
Ubiquitination		↑	1
Unknown		↑	1

Application of gene expression profiling through microarray in segregating population for sucrose accumulation capacity has allowed identification of genes for this important trait in sugarcane (Casu et al. 2005; Felix et al. 2009; Papini-Terzi et al. 2009). A putative serine-threonine protein phosphatase encoding gene and an unknown gene with altered expression in high and low sucrose segregants were reported by Casu et al. (2005). Twenty four genes were reported to be expressed differentially in high and low sugar progeny, with 19 having a higher level of expression in low sugar genotypes (Felix et al. 2009) (Table 2). Evidence of cross-talk between hormone biosynthesis exists in this study as pointed out by other workers also (Vicentini et al. 2009). The over expression of a large number of stress related genes in low sugar content plants indicates that the sugar accumulation pathway is very complex with large scale interactions taking place between these pathways The activation of signal transduction components appear to be a long term mechanism, rather than being a one-time strategy.

The complexity of sucrose accumulation pathway and its interaction with other metabolic processes like stress regulation was observed by Papini-Terzi et al. (2009) also in their investigation of 30 sugarcane genotypes with varying Brix, using cDNA microarrays. Protein phosphorylation was found to have an association with sucrose accumulation and culm development. A predominance of stress related gene families among the genes associated with sugar content was reported here also. A limited overlap with ABA signalling was observed. Several protein kinases and transcription factors may act as regulators of sucrose accumulation. Aquaporins, lignin biosynthesis genes and cell wall synthesis genes were other categories of genes found to be associated with sucrose accumulation. This possible overlap among these different pathways also suggests the possibility of utilization of some of these gene sequences as molecular markers for sucrose content in breeding programmes.

In spite of a very strong overlap among the various pathways, the regulatory mechanisms may vary in different situations as evidenced in the case of drought response (Iskandar et al. 2011). A sub-set of stress related genes identified (that were potentially associated with sucrose content in sugarcane) were analyzed for their expression and regulation. It was speculated that stress response genes may be activated in sugarcane culm as a protective mechanism against the high solute concentration in storage cells and associated osmotic gradients developed. The stress tolerance mechanism that facilitates normal cellular functioning even under high sugar load may get activated by sucrose accumulation. The genes activated under such a situation may have similarities with those genes that are activated under normal stress conditions. But the protection mechanism seems to be different in the two situations (Iskander et al. 2011). Here also, the role of the putative sugar transporter gene *PST5* in regulation of sucrose accumulation in sugarcane was reinforced.

5.6.1 Molecular Marker Application Through Reverse Genetics Approaches

Reverse genetics strategies for improving sugar content also involve the application of molecular markers, if not directly. As mentioned earlier, the sequences identified through marker application can be used for transformation or the information generated through reverse genetics can form a basis for molecular marker development. The usual methods of reverse genetics involve (i) development of mutant population (ii) gene silencing for reduced expression of target genes and (iii) over expression of target gene in the plant. Due to the highly polyploid nature of the plant, mutation studies may not be effective in sugarcane for expression analysis. Since sugarcane is highly amenable to transformation and regeneration (Bower and Birch 1992; Basnayake et al. 2011; Joyce et al. 2010) the use of gene silencing and over expression studies hold great promise. Very few studies have used these methods to study expression of specific genes in various physiological processes in sugarcane.

Transgenic plants exhibiting alteration in sucrose levels have also helped in understanding the sugar metabolic pathways to some extent. High yields of the high value sugar isomaltulose (IM) was obtained by vacuole targeting of an efficient sucrose isomerise (SI) without disrupting the growth and development of the plant, through a transgenic approach (Wu and Birch 2007). This resulted in doubling of total sugar content in mature internodes of an elite high sugar variety overcoming the constraints of osmotic limits and osmotic constraints as has been the belief before (Moore 1995; Jackson 2005). This in turn, opened up the possibility of increased sugar: water ratio rather than increased sugar:fibre partitioning for enhanced sugar content. But the recent field trials involving these transformants indicated a comparable decrease in sucrose concentration in these plants, with no overall decrease in total sugars. Thus there is a need for further research in this direction (Basnayake et al. 2012). Reverse genetics analysis of the function of pyrophosphate:fructose-6-phosphate 1-transferase (PFP) is a classic example where the function of a gene has been dissected to reveal its stage-specific role in sugar accumulation in sugarcane (Groenewald and Botha 2008;

van der Merwe et al. 2010). Whittaker and Botha (1999) had demonstrated a correlative association for this gene with sugar accumulation. Down regulation of PFP led to an increase in hexose phosphate:triose phosphate ratio in immature internodes there by leading to an increase in sucrose synthesis in immature internodes. Gluconeogenesis was also found to increase in mature internodes. Suppression of neutral invertase through antisense technology led to increased sucrose content in sugarcane suspension callus. The regenerants (transgenic plants) exhibited an increased sucrose: hexose ration and reduced sucrose recycling in the culm (Rossouw et al. 2010). The essential role of *PST2a* gene, an important sugar transporter in sugarcane culm in plant development as well as sugar accumulation was demonstrated when RNAi strategy was tried for suppression of gene expression (Casu et al. 2003). The low number of transgenic regenerants and absence of reliable down regulation of this transcript proved this point.

Sugar manipulation by engineering a new carbon sink into sugarcane in the form of sorbitol (Chong et al. 2007) was also attempted. Diverting hexose phosphate equivalents in sugarcane to an alternative carbon sink resulted in increased expression of sucrose synthesizing and cleavage enzymes, without effecting sink sucrose accumulation. This also suggested that sugarcane metabolic intermaediates that are normally used for sucrose synthesis may be diverted to alternate sink without much adverse effects.

High Throughput Array Differential Screening revealed that promoters for genes *DIRIGENT* (SHD1R16) and *O-Methyl Transferase* (SHOMT) conferred stem regulated gene expression (Damaj et al. 2010). This points towards the possibility of uitlizing these in carbon metabolism and bioenergy production studies. The specificity of their expression in the stem, in particular in the vascular bundles, and enhancement of their expression by stress regulators make these promoters very valuable for metabolic engineering through sugar accumulation or fibre content as the necessity demands. These may play a key role in carbon partitioning.

Though these studies may not directly involve molecular marker utilization, the information generated can be successfully put to use in molecular marker studies. Conversely, candidate gens identified through marker applications can be utilized in these reverse genetics studies thereby aiding in sucrose manipulation in the sugarcane culms.

5.7 Molecular Ctyogenetics for Sugar Trait

Even though molecular cytogenetic techniques like genomic in situ hybridization (GISH) and fluorescence in situ hybridization (FISH) have made significant contribution towards sugarcane improvement as mentioned before (D'Hont et al. 1995, 1996), no such studies have been reported where molecular cytogenetic probes have been put to use in diversity studies, mapping or other such studies related to sugar trait. But these investigations have been the basis of valuable information that had been put to use in linkage mapping and QTL mapping studies in this crop. Piperidis et al. (2010) in their investigations on chromosome composition and transmission in modern cultivars have confirmed that, on an average, 70–80% of the genome was constituted by *S. officinarum*, 10–23% by *S. spontaneum* complete chromosomes and 8–13% by interspecific exchanges (Piperidis et al. 2010). These exchanges show progressive accumulation in subsequent generations. The occurrence of 2n + n transmission in *S. officinarum* × modern cultivars crosses also was demonstrated in this study. These findings have important implications in molecular marker applications like linkage and QTL mapping in this crop.

As in the case of other DNA based markers, sequences specific to sugar trait once identified can be used as marker probes and these can be used to locate the sugar related sequences in the genome of cultivated clones. The transmission of these sequences in the progeny can also be studied thus, aiding in selection of desirable progeny with target sequences. Identification of molecular cytogenetic probes for sugar trait can be of help in further mapping studies, comparative genomics and also in the functional genomic studies. Thus this is one area of research that has great potential.

6 Molecular Markers in Sugarcane for Sugar Traits: Where Do We Go from Here?

Having reviewed the various reports of molecular marker applications in sugarcane for sugar and related traits, where do all these results lead us to? Undoubtedly, the high scope that this area of research offers, in further understanding the complex pathway of sucrose metabolism is unparallelled.

The information revealed through expression analyses has helped us to identify the role of several key genes in sugar accumulation. The role of source-sink regulation in sugar accumulation, the possibility of uncoupling of sink to source signals, the futile cycle taking place in the culm etc., in regulating the final sucrose accumulation are being investigated with renewed interest at a molecular level. Thus, in addition to the usual enzymes like SPS, sucrose synthase and invertases, sequences coding for sucrose transporters, cell membrane permeability genes, carbon partitioning genes or even genes for sucrose signalling and transcription factors may be potential candidates for use in molecular marker identification studies. The interaction with genes involved in other pathways makes those genes also probable players. The role of alternative splicing and other regulatory mechanisms also need to be addressed in detail. The difference in repetitive sequences in varieties may have a role in their varying adaptability. This may also play a role in the variation with respect to sucrose accumulation. This also forms a potent area of research for the future. The genes that result in the peculiar Kranz anatomy in these C_4 leaves, those involved in the spatial separation of the various processes etc., also need to be identified to study the specific role of this structural peculiarity in the sugar accumulation capacity of the crop. Mechanisms of genomic buffering, repeat sequence restriction in sugarcane compared to sorghum and its relation to sugar accumulation etc. are other important questions to be answered making use of molecular markers. The finding that only single dose of some multi-dose loci get expressed in sugarcane, points to some sort of regulatory mechanism operating at this level, which opens up a possible role of this mechanism of regulation in sugar accumulation also. Selection of low copy number sequences and euchromatin sequences using comparative genomics will provide a reasonable starting point for such molecular marker related studies. Using pollen to isolate DNA/RNA may also result in less copy number of genes to be handled.

Studies till now undermine the fact that strategies like marker identification, mapping of trait linked loci etc. alone may not be sufficient for proper understanding of sugar accumulation. Thus other areas like expression studies also need to be explored. MicroRNA (miRNA) based regulation is a potential area of interest in sugarcane as indicated by Zanca et al. (2010). miRNAs are small regulatory RNAs that play a role in various physiological and metabolic pathways like stress response, signal transduction, protein degradation etc. This miRNA associated RNA silencing leading to translational inhibition, mRNA decay/cleavage can regulate different crucial steps in sugar metabolic pathway also. Another possibility can be to utilize these miRNAs for reducing the level of inversion that may take place in some early maturing high sugar varieties at later stages of the season. Such early varieties which show a decline in sugar content during the later period, can be potential candidates for miRNA induced silencing studies, thereby sustaining their high sugar content throughout the period. Since several of the miRNAs show homology with sorghum miRNAs these can be of use in comparative genomics also. The possibility that some of the MITE-associated miRNAs may lose their miRNA status over time, necessitates a cautious approach in this regard.

The evolution of molecular breeding has led to a deeper understanding of the underlying trait-linked loci and their interactions. Gradually this is progressing towards quantification of transcript levels or expression QTLs (eQTLS). Such genetical genomic studies in the long run, will lead to elucidation of the entire genome, and thereby the different biochemical pathways. Serial Analysis of Gene Expression (SAGE), Massive Parallel Signature Sequencing (MPSS) etc. are some of the techniques that can be employed. Though an initiation has been made in sugarcane, large scale eQTLs studies neeed to be taken up in this crop. There is every possibility that genes with no previously-assigned functions may act as master regulators thereby, facilitating their incorporation at specific points in the sugar metabolic pathway. Once allelic variations are linked with phenotypic variations these cloned regulatory genes can be put to use in molecular breeding studies. Thus QTL mapping combined with eQTLs can aid in identifying biomarkers for classical phenotypes. A word of caution is needed here also. In polyploids like sugarcane, the genome buffering may mask the functioning of some negative genes, which may be problematic. Also the low level of resolution in most of the presently available genetic maps in this crop may be a drawback for some applications which may be overcome in the long run. Needless to say all these techniques should go hand-in-hand with proper phenotyping studies with sufficient sample size, replications, locations etc., in the case of field trials and validation studies.

With all the techniques available, the utilization of proper bioinformatic tools and software is very important for the successful implementation. The large scale sequencing initiatives need to have excellent collaboration with the different sugarcane research workers all over the world so that these initiatives can lead to fruitful conclusions.

References

Aitken KS, Jackson PA, McIntyre CL (2005) A combination of AFLP and SSR markers provides extensive map coverage and identification of homo(eo)logous linkage groups in a sugarcane cultivar. Theor Appl Genet 110:789–801

Aitken KS, Jackson PA, McIntyre CL (2006) Quantitative trait loci identified for sugar related traits in sugarcane (*Saccharum* spp.) cultivar × *Saccharum officinarum* population. Theor Appl Genet 112:1306–1311

Aitken KS, Jackson PA, McIntyre CL (2007) Construction of genetic linkage map of *Saccharum officinarum* incorporating both simplex and duplex markers to increase genome coverage. Genome 50:742–756

Aitken KS, Hermann S, Karno K, Bonnett GD, McIntyre LC, Jackson PA (2008) Genetic control of yield related stalk traits in sugarcane. Theor Appl Genet 117:1191–1203

Al-Janabi SM, Honeycutt RJ, McClelland M, Sobral BWS (1993) A genetic linkage map of *Saccharum spontaneum* L. "SES 208". Genetics 134:1249–1260

Al-Janabi SM, McClelland M, Petersen C, Sobral BWS (1994) Phylogenetic analysis of organellar DNA sequences in the Andropogoneae: Saccharinae. Theor Appl Genet 88:933–944

Al-Janabi SM, Parmessur Y, Kross H, Dhayan S, Saumtally S, Ramdoyal K, Autrey LJC, Dookun -Saumtally A (2007) Identification of a major quantitative trait locus (QTL) for yellow spot (*Mycovellosiella koepkei*) disease resistance in sugarcane. Mol Breeding 19:1–14

Alwala S, Suman A, Arro JA, Veremis JC, Kimberg CA (2006a) Target region amplification polymorphism (TRAP) for assessing genetic diversity in sugarcane germplasm collections. Crop Sci 46:448–455

Alwala S, Kimberg CA, Gravois KA, Bischoff KP (2006b) Trap, a new tool for sugarcane breeding: comparison with AFLP and co-efficient of parentage. Sugarcane Int 24(6):11–21

Alwala S, Collins A, Kimbeng J, Veremis C, Gravois KA (2008) Linkage mapping and genome analysis in a *Saccharum* interspecific cross using AFLP, SRAP and TRAP markers. Euphytica 164:37–51

Alwala S, Collins A, Kimbeng J, Veremis C, Gravois KA (2009) Identification of molecular markers associated with sugar-related traits in a *Saccharum* interspecific cross. Euphytica 167:127–142

Asnagahi C, Roques D, Ruffel S, Kaye C, Hoarau JY, Telismart H, Girard JC, Raboin LM, Risterucci AM, Grivet L, D'Hont A (2004) Targeted mapping of a sugarcane rust resistance gene (Bru1) using bulked segregant analysis and AFLP markers. Theor Appl Genet 108:759–764

Basnayake SWV, Moyle R, Birch G (2011) Embryogenic callus proliferation and regeneration conditions for genetic transformation of diverse sugarcane cultivars. Plant Cell Rep 30(3):439–448

Basanayake SWV, Morgan CT, Wu L, Birch RG (2012) Field performance of transgenic sugarcane expressing isomaltulose synthase. Plant Biotechnol J 10(2):217–225

Besnard G, Offmann B, Robert C, Rouch C, Cadet F (2002) Assessment of the C_4 phosphoenolpyruvate carboxylase gene diversity in grasses (Poaceae). Theor Appl Genet 105:404–412

Bhatt SR, Gill SS (1985) The implications of 2n egg gametes in nobilization and breeding of sugarcane. Euphytica 34:377–384

Bodenes C, Laigret F, Kremer A (1996) Inheritance and molecular variations of PCR-SSCP fragments in pedunculate oak (*Quercus robur* L.). Theor Appl Genet 93:348–354

Bonierbale MW, Plaisted RL, Tanksley SD (1988) RFLP maps based on a common set of clones reveal modes of chromosomal evolution in potato and tomato. Genetics 120:1095–1103

Botstein D, White RL, Stocknick M, Davis RW (1980) Construction of linkage map using restriction fragment length polymorphism. Am J Hunam Genet 32:314–331

Bower R, Birch RG (1992) Transgenic sugarcane plant via microprojetcile bombardment. Plant J 2:409–416

Bowyer JR, Leegood RC (1997) Photosynthesis. In: Day PM, Harbone JB (eds) Plant biochemistry. Academic Press, San Diego, pp 49–110

Bremer G (1923) A cytological investigation of some species and species hybrids within the genus *Saccharum*. Genetica (The Hague) 5(97–148):273–326

Bremer G (1924) The cytology of sugarcane. A cytological investigation of some cultivated kinds and their parents. Genetica (The Hague) 6:497–525

Bremer G (1925) The cytology of sugarcane. The chromosomes of primitive forms of the genus *Saccharum*. Genetica (The Hague) 7:293–322

Bremer G (1961a) Problems in breeding and cytology of sugarcane. I. A short history of sugarcane breeding—the original forms of *Saccharum*. Euphytica 10:59–78

Bremer G (1961b) Problems in breeding and cytology of sugarcane. II. The sugarcane breeding from a cytological viewpoint. Euphytica 10:121–133

Bremer G (1961c) Problems in breeding and cytology of sugarcane. III. The cytological crossing research of sugarcane. Euphytica 10:229–243

Bremer G (1961d) Problems in breeding and cytology of sugarcane. IV. The origin of increase in chromosome number in species hybrids *Saccharum*. Euphytica 10:325–342

Bremer G (1962) Problems in breeding and cytology of sugarcane. V. Chromosome increase in *Saccharum* hybrids in relation to interspecific and intergeneric hybrids in other genera. Euphytica 11:65–80

Brown AHD, Daniels J, Latter BDH (1968) Quantitative genetics of sugarcane. I. Analysis of variation in a commercial hybrid sugarcane population. Theor Appl Genet 38:361–369

Burnquist WL, Sorrells ME, Tanksley S (1992) Characterization of genetic variability in *Saccharum* germplasm by means of restriction fragment length polymorphism (RFLP) analysis. Proc Int Soc Sugarcane Technol 21:355–365

Calsa T Jr, Figuera A (2007) Serial analysis of gene expression in sugarcane (*Saccharum* spp.) leaves revealed alternative C_4 metabolism and putative antisense transcripts. Plant Mol Biol 63:745–762

Carson DL, Huckett BI, Botha FC (2002) Sugarcane ESTs differentially expressed in immature and maturing intermodal tissue. Plant Sci 162:289–300

Casu RE, Grof CP, Rae AL, McIntyre CL, Dimmock CM, Manners JM (2003) Identification of a novel sugar transporter homologue strongly expressed in maturing stem vascular bundle tissues of sugarcane by expressed sequence tag and microarray analysis. Plant Mol Biol 52(2):371–386

Casu RE, Dimmock GM, Chapman C, Grof CP, McIntyre CL, Bonnett GD, Manners JM (2004) Identification of differentially expressed transcripts from maturing stem of sugarcane by in silico analysis of stem expressed sequence tags and gene expression profiling. Plant Mol Biol 54(4):503–517

Casu RE, Manners JM, Bonnett GD, Jackson PA, McIntyre CL, Dunne R, Chapman SC, Rae AL, Grof CPL (2005) Genomics approaches for identification of genes determining important traits in sugarcane. Field Crop Res 92:137–147

Causse M, Chaib J, Lecomte L, Buret M, Hospital F (2007) Both additive and epistasis control the genetic variation for fruit quality traits in tomato. Theor Appl Genet 115(3):429–442

Chaib J, Lecomte L, Buret M, Causse M (2006) Stability over genetic backgrounds, generations and years of quantitative trait locus (QTLs) for organoleptic quality in tomato. Theor Appl Genet 112(5):934–944

Chandra A, Jain R, Rai RK, Solomon S (2011) Revisiting the source-sink paradigm in sugarcane. Curr Sci 100(7):978–980

Charcosset A, Moreau L (2004) Use of molecular markers for the development of new cultivars and the evaluation of genetic diversity. Euphytica 37:81–94

Cho YG, Ishii T, Temnykh S, Chen X, Lipovich L, McCouch SR, Park WD, Ayres N, Cartinhour S (2000) Diversity of microsatellites derived from genomic libraries and GenBank sequences in rice (*Oryza sativa* L.). Theor Appl Genet 100:713–722

Chong BF, Bonnett GD, Glassop D, O'Shea MG, Brumbley SM (2007) Growth and metabolism in sugarcane are altered by the creation of a new hexose-phosphate sink. Plant Biotechnol J 5:240–253

Collard BCY, Jahufer MZZ, Brouwer JB, Pang ECK (2005) An introduction to markers, quantitative trait loci (QTL) mapping and marker-assisted selection for crop improvement: the basic concepts. Euphytica 142:169–196

Collard BCY, Mackill DJ (2008) Marker-assisted selection: an approach for precision plant breeding in the twenty-first century. Phil Trans R Soc B 363:557–572

Cordeiro GM, Maguire TL, Edwards KJ, Henry RJ (1999) Optimisation of a microsatellite enrichment technique in *Saccharum* spp. Plant Mol Biol Rep 17:225–229

Cordeiro GM, Taylor GO, Henry RJ (2000) Characterization of microsatellite markers from sugarcane (*Saccharum* spp.) a highly polyploidy species. Plant Sci 155:161–168

Cordeiro GM, Casu R, McIntyre CL, Manners JM, Henry RJ (2001) Microsatellite markers from sugarcane (*Saccharum* spp.): ESTs cross-transferable to erianthus and sorghum. Plant Sci 160:1115–1123

D'Hont A, Lu YH, Feldmann P, Glaszmann JC (1993) Cytoplasmic diversity in sugarcane revealed by heterologous probes. Sugar Cane 1:12–15

D'Hont A, Lu YH, Le'on DGD, Grivet L, Feldmann P, Panaud E, Glaszmann JC (1994) A molecular approach to unravelling the genetics of sugarcane, a complex polyploid of Andropogonaea tribe. Genome 37:222–230

D'Hont A, Grivet L, Feldmann P, Rao PS, Berding N, Glaszmann JC (1995) Identification and characterization of sugarcane intergeneric hybrids, *Saccharum officinarum* × *Erianthus arundinaceus* with molecular markers and DNA in situ hybridization. Theor Appl Genet 91:320–326

D'Hont A, Grivet L, Feldmann P, Rao PS, Berding N, Glaszmann JC (1996) Characterisation of the double genome structure of modern sugarcane cultivars (*Saccharum* spp.) by molecular cytogenetics. Mol Gen Genet 250:405–413

D'Hont A, Ison D, Alix K, Roux C, Glaszmann JC (1998) Determination of basic chromosome number in the genus *Saccharum* by physical mapping of ribosomal RNA genes. Genome 41:221–225

Da Silva JA, Bressiani JA (2005) Sucrose synthase molecular marker associated with sugar content in elite sugarcane progeny. Genet Mol Bio 1 28:294–298

Da Silva JAG, Sorrells ME, Burnquist W, Tanksley SD (1993) RFLP linkage map and genome analysis of *Saccharum spontaneum*. Genome 36:782–791

Da Silva J, Honeycutt RJ, Burnquist W, Al-Janabi SM, Sorrells ME, Tanksley SD, Sobral BWS (1995) *Saccharum spontaneum* L. 'SES 208'' genetic linkage map combining RFLP and PCR-based markers. Mol Breeding 1:165–179

Damaj MB, Kumpatla SP, Emani C, Beremand PD, Reddy AS, Rathore KS, Buenrostro-Nava MT, Curtis IS, Thomas TL, Mirkov TE (2010) Sugarcane DIRIGENT and O-METHYL-TRANSFERASE promoters confer stem-regulated gene expression in the diverse monocots. Planta 231:1439–1458

Daniels J, Roach BT (1987) Taxonomy and evolution. In: Heinz DJ (ed) Sugarcane improvement through breeding. Elsevier, Amsterdam, pp 7–84

Daugrois JH, Grivet L, Roques D, Hoarau JY, Lombard H, Glaszimann JC, Hont AD (1996) A putative major gene for rust resistance linked with an RFLP marker in sugarcane cultivar R 570. Theor Appl Genet 92:1059–1064

Dufour P, Grivet L, D'Hont A, Deer M, Tronche G, Glaszmann JC, Hamon P (1996) Comparative genetic mapping between duplicated segments on maize chromosome 3 and 8 and homologous regions in *Sorghum* and sugarcane. Theor Appl Genet 92:1024–1030

Dufour P, Deu M, Grivet L, D'Hont A, Paulet F, Bouet A, Lanaud C, Glaszmann JC, Hamon P (1997) Construction of a composite sorghum genome map and comparison with sugarcane, a related complex polyploid. Theor Appl Genet 94:409–418

Dutt NL, Rao KSS (1933) Observations on the cytology of sugarcane. Indian J Agric Sci 3:37–56

Dutt NL, Rao JT (1951) The present taxonomic position of *Saccharum* and its congeners. In: Proceedings of 7th ISSCT, Brisbane, pp 228–293

Edme SJ, Glynn NG, Comstock JC (2006) Genetic segregation of microsatellite markers in *Saccharum officinarum* and *S. spontaneum*. Heredity 97:366–375

Eshed Y, Zamir D (1996) Less-than-additive epistatic interactions of quantitative trait loci in tomato. Genetics 143:1807–1817

Felix JM, Papini-Terzi FS, Rocha FR, Vencio RZN, Vicentini R, Nishiyama Jr MY, Ulian EC, Souza GM, Menossi M (2009) Expression profile of signal transduction components in a sugarcane population segregating for sugar content. Trop Plant Biol 2: 98–109

Fukuoka S, Inoue T, Miyao A, Monna L, Zhong HS, Sasaki T, Minobe Y (1994) Mapping of sequence-tagged sites in rice by single strand conformation polymorphism. DNA Res 1:271–277

Garcia AAF, Kido EA, Meza AN, Souza HMB, Pinto LR, Pastina MM, Leite CS, da Silva JAG, Ulian EC, Figueira A, Souza AP (2006) Development of an integrated genetic map of a sugarcane (*Saccharum* spp.) commercial cross, based on a maximum likelihood approach for estimation of linkage and linkage phases. Theor Appl Genet 112:298–314

Glassop D, Roessner U, Bacic A, Bonnett GD (2007) Changes in sugarcane metabolome with stem development. Are they related to sucrose accumulation? Plant Cell Physiol 48(4):573–584

Glaszmann JC, Fautret A, Noyer JL, Feldmann P, Lanaud C (1989) Biochemical genetic markers in sugarcane. Theor Appl Genet 78:537–543

Glaszmann JC, Lu YH, Lanaud C (1990) Variation of nuclear ribosomal DNA in sugarcane. J Genet Breed 44:191–198

Glaszmann JC, Dufour P, Grivet L, D'Hont A, Deu M, Paulet F, Hamon P (1997) Comparative genome analysis between several tropical grasses. Euphytica 96:13–21

Grivet L, Arruda P (2001) Sugarcane genomics: depicting the complex genome of an important tropical crop. Curr Opin Plant Biol 5:122–127

Grivet L, D'Hont A, Dufour P, Hamon P, Roquest D, Glaszmann JC (1994) Comparative genome mapping of sugarcane with other species within the Andropogoneae tribe. Heredity 73:500–508

Grivet L, D'Hont A, Roques D, Feldmann P, Lanaud C, Glaszmann JC (1996) RFLP mapping in cultivated sugarcane (*Saccharum* spp.): genome organization in a highly polyploid and aneuploids interspecific hybrid. Genetics 142:987–1000

Grivet L, Glaszmann JC, Arruda P (2001) Sequence polymorphism from EST data in sugarcane: a fine analysis of 6-phosphogluconate dehydrogenase. Genet Mol Biol 24(1–4):161–167

Groenewald JH, Botha FC (2008) Down regulation of pyrophosphate:D-fructose-6-phosphate 1-phosphotransferase activity in sugarcane enhances sucrose accumulation in immature internodes. Transgenic Res 17:85–92

Guimaraes CT, Sills GR, Sobral BWS (1997) Comparative mapping of Andropogoneae: *Saccharum* L. (sugarcane) and its relation to sorghum and maize. Proc Natl Acad Sci (USA) 94:14261–14266

Guimaraes CT, Honeycutt RJ, Sills GR, Sobral BWS (1999) Genetic maps of *Saccharum officinarum* L. and *Saccharum robustum* Brandes & Jew. Ex Grassl. Genet Mol Biol 22:125–132

Gupta PK, Balyan HS, Sharma PC, Ramesh B (1996) Microsatellites in plants: a new class of molecular markers. Curr Sci 70:45–54

Harvey H, Huckett BI, Botha FC (1994) Use of polymerase chain reaction and random amplification of polymorphic DNAs for the determination of genetic distances between 21 sugarcane varieties. In: Proceedings of South African Sugar Technology Association, June, pp 36–40

Hatch MD (1977) C_4 pathway photosynthesis: Mechanism and physiological function. Trends Biochem Sci 2:199–201

Hatch MD, Glasziou KT (1963) Sugar accumulation cycle in sugarcane II. Relationship of invertase activity to sugar content and growth rate in storage tissue of plants grown in controlled environments. Plant Physiol 38:344–348

Hatch MD, Sacher JA, Glasziou KT (1963) Sugar accumulation cycle in sugarcane I. Studies on enzymes of the cycle. Plant Physiol 38:338–343

Heller-Uszynska K, Uszynska G, Huttner E, Evers M, Carlig J, Caig V, Aitken K, Jackson P, Piperidis G, Cox M, Gilmour R, D'Hont A, Butterfield M, Glaszmann JC, Kilian A (2011) Diversity array technology effectively reveals DNA polymorphism in a large and complex genome of sugarcane. Mol Breed 28:37–55

Hemaprabha G, Natarajan US, Balasundaram N, Singh NK (2006) STMS based genetic divergence among common parents and its use in identifying productive cross combinations for varietal evolution in sugarcane (*Saccharum* sp.) sugarcane. International 24(6):22–27

Herbert LP, Henderson MT (1959) Breeding behaviour of certain agronomic characters in progenies of sugarcane crosses. USDA Tech Bull. No 1195:54

Hoarau JY, Offmann B, D'Hont A, Risterucci AM, Roques D, Glaszmann JC, Grivet L (2001) Genetic dissection of a modern sugarcane cultivar (*Saccharum* spp.) I. Genome mapping with AFLP markers. Theor Appl Genet 103:84–97

Hoarau JY, Grivet L, Offmann B, Raboin LM, Diorflar JP, Payet J, Hellmann M, D'Hont A, Glaszmann JC (2002) Genetic dissection of a modern sugarcane cultivar (*Saccharum* spp.) II. Detection of QTLs for yield components. Theor Appl Genet 105:1027–1037

Hogarth DM (1968) A review of quantitative genetics in plant breeding with particular reference to sugarcane. J Aust Inst Agric Sci 34:108–120

Hogarth DM, Wu KK, Heinz DJ (1981) Estimating genetic variance in sugarcane using a factorial cross design. Crop Sci 21:21–25

Hospital F (2009) Challenges for effective marker assisted selection in plants. Genetica 136:303–310

Hsu SY, Hour AL, Wang TH (1995) Heritability and modes of inheritance of brix in sugarcane seedlings. Proc ISSCT 22:286–292

Huang JW, Chen JT, Yu WP, Shyur LF, Wang AY, Sung HY, Lee PD, Su JC (1996) Complete structures of three rice sucrose synthase isogenes and differential regulation of their expressions. Biosci Biotech Biochem 60:233–239

Hulbert SH, Rcheter TE, Axtell JD, Bennetzen JL (1990) Genetic mapping and characterization of sorghum and related crops by means of maize DNA probes. Proc Natl Acad Sci (USA) 87:4251–4255

Iskandar HM, Casu RE, Fletcher AT, Scmidt S, Xu J, Maclean D, Manners JM, Bonnett GD (2011) Identification of drought-response genes and a study of their expression during sucrose accumulation and water deficit in sugarcane culms. BMC Plant Biology 11:12 doi:10.1186/1471-2229-11-12

Jackson PA (2005) Breeding for improved sugar content in sugarcane. Field Crops Res 92:277–290

Jagathesan D, Sreenivasan TV (1967) Cytogenetical studies in *Narenga porphyrocoma* I. Study of karyotype and abnormalities in meiosis. Cytologia 32:11–18

Jagathesan D, Ramadevi KN (1969) Cytogenetical studies in *Erianthus* I Chromosome morphology in 2n = 20 forms. Caryologia 22(4):359–368

Jagathesan D, Sreenivasan TV (1971) Cytogenetical studies in *Erianthus*. Occurrence of chromosome knobs in 2n = 20 forms. Caryologia 24(1):27–31

Jalaja NC (1983) Cytogenetical studies in *Saccharum* and allied genera. Aneuploids in *Saccharum spontaneum* L. Annual report of Sugarcane Breeding Institute, Coimbatore

Janaki Ammal EK (1938a) A *Saccharum-Zea* cross. Nature (Lond.) 142:618–619

Janaki Ammal EK (1938b) Chromosome behavior in *S.spontneum* × *Sorghum durra* hybrid. In: Proceedings of Indian Science Congress Association, part 3, p 143

Janaki Ammal EK, Jagathesan D, Sreenivasan TV (1972) Further studies in *Saccharum-Zea* hybrid I. Mitotic Studies Heredity 28:141–142

Jannoo N, Grivet L, Dookun A, D'Hont A, Glaszmann JC (1999) Linkage disequilibrium among modern sugarcane cultivars. Theor Appl Genet 99:1053–1060

Jena KK, Khush GS (1989) Monosomic alien addition lines of rice: production, morphology, cytology and breeding behaviour. Genome 32(3):449–455

Joyce P, Kuwahata M, Turner N, Lakshmanan P (2010) Selection systems and co-cultivation medium are important determinants of *Agrobacterium* mediated transformation of sugarcane. Plant Cell Rep 29:173–183

Kam-Morgan LNW, Gill BS (1989) DNA restriction fragment length polymorphism: a strategy for genetic mapping of D genome of wheat. Genome 32:724–732

Kearsay MJ (1998) The principles of QTL analysis (a minimal mathematics approach). J Exp Bot 49:1619–1623

Khan MS, Yadav S, Srivastava S, Swapna M, Chandra A, Singh RK (2011) Development and utilization of conserved intron-scanning primers in sugarcane. Australian J Bot. 59(1):38–45

Kortschak HP, Hartt CE, Burr GO (1965) Carbon dioxide fixation in sugarcane leaves. Plant Physiol 40:209–213

Laetsch WM (1974) The C_4 syndrome. A structural analysis. Ann Rev Plant Physiol 25:27–52

Lecomte L, Duffe P, Buret M, Servin B, Hospital F, Causse M (2004) Marker-assisted introgression of five QTLs controlling fruit quality traits into three tomato lines revealed interactions between QTLs and genetic backgrounds. Theor Appl Genet 109(3):658–668

Liao CY, Wu PH, Hu B, Yi KK (2001) Effects of genetic background and environment on QTLs and epistasis of rice (*Oryza sativa* L.) panicle number. Theor Appl Genet 108:141–153

Lima MLA, Garcia KM, Oliveira KM, Matsuoka S, Arizono H, De Souza CL Jr, de Souza AP (2002) Analysis of genetic similarity detected by AFLP and co-efficient of parentage among genotypes of sugarcane (*Saccharum* spp.). Theor Appl Genet 104:30–38

Lingle SE (1996) Rate of growth and sugar accumulation in sugarcane related to sucrose synthase activity. In: Wilson JR, Hogarth DM, Campbell JA, Garside AL (eds) Sugarcane: research towards efficient and sustainable production. CSIRO Division of Tropical Crops and Pastures, Brisbane, pp 95–97. Proceedings of Sugar 2000 Symposium, 19–23 August 1996, Brisbane, Australia

Lingle SE (1997) Seasonal internode development and sugar metabolism in sugarcane. Crop Sci 37:1222–1127

Lingle SE (1999) Sugar metabolism during growth and development in sugarcane internodes. Crop Sci 39:480–486

Lingle S, Dyer JM (2004) Polymorphism in the promoter region of the sucrose synthase-2 gene of *Saccharum* genotypes. J Am Soc Sugar Cane Technol 24:241–249

Liu BH (1998) Statistical genomics. CRC Press, New York 611 pp

Liu P, Que Y, Pan YB (2011) Highly polymorphic microsatellite DNA markers for sugarcane germplasm evaluation and variety identity testing. Sugar Tech 13(2):129–136

Lu YH, D'Hont A, Walker DJT, Rao PS, Feldmann P, Glaszmann JC (1994) Relationship among ancestral species of sugarcane revealed with RFLP using single copy maize nuclear probes. Euphytica 78:7–18

Manners JM, Casu RE (2011) Transcriptome analysis and functional genomics of sugarcane. Trop Plant Biol 4:9–21

McClintock B, Hill HE (1931) The cytological association of chromosomes associated with R-G linkage group in *Zea* mays. Genetics 16:175–190

McCormick AJ, Cramer MD, Watt DA (2008a) Differential expression of genes in the leaves of sugarcane in response to sugar accumulation. Trop Plant Biol 1:142–158

McCormick AJ, Cramer MD, Watt DA (2008b) Regulation of photosynthesis by sugars in sugarcane leaves. J Plant Physiol 165: 1817–1829

McCormick AJ, Watt DA, Cramer MD (2009) Supply and demand: sink regulation of sugar accumulation in sugarcane. J Exp Bot 60(2):357–364

McIntyre CL, Jackson M, Cordeiro GM, Amouyal O, Hermann S, Aitken KS, Eliott F, Henry RJ, Casu RE, Bonnett GD (2006) The identification and characterization of alleles of sucrose phosphate synthase gene family III in sugarcane. Mol Breed 18:39–50

Ming R, Liu SC, Lin YR, Da Silva J, Wilson W, Braga D, Van Deynze A, Wenslaffe TE, Wu KK, Moore PH, Burnquist W, Sorrells ME, Irvine JE, Paterson AH (1998) Detailed alignment of *Saccharum* and sorghum chromosomes: comparative organization of closely related diploid and polyploid genomes. Genetics 150:1663–1682

Ming R, Liu SC, Moore PH, Irvine JE, Paterson AH (2001) QTL analysis in a complex autopolyploid: genetic control of sugar content in sugarcane. Genomic Res 11:2075–2084

Ming R, Wang YW, Draye X, Moore PH, Irvine JE, Paterson AH (2002) Molecular dissection of complex traits in autopolyploids: mapping QTLs affecting sugar yield and related traits in sugarcane. Theor Appl Genet 105:332–345

Ming R, Moore PH, Wu KK, D'Hont A, Glaszmann JC, Tew TL (2006) Sugarcane improvement through breeding and biotechnology. Plant Breed Rev 27:15–118

Moore PH (1995) Temporal and spatial regulation of sucrose accumulation in the sugarcane stem Australian. J Plant Physiol 22:661–679

Mudge J, Andersen WR, Kehrer RL, Fairbanks DJ (1996) A RAPD genetic map of *Saccharum officinarum*. Crop Sci 36:1362–1366

Mukherjee SK (1957) Origin and distribution of *Saccharum*. Bot Gaz 119:55–61

Nair NV, Nair S, Sreenivasan TV, Mohan M (1999) Analysis of genetic diversity and phylogeny in *Saccharum* and related genera using RAPD markers. Genet Resour Crop Evol 46:73–79

Nair NV, Selvi A, Sreenivasan TV, Pushpalatha KN (2002) Molecular diversity in Indian sugarcane cultivars as revealed by random amplified DNA polymorphisms. Euphytica 127:219–225

Nair NV, Selvi A, Sreenivasan TV, Pushpalatha KN, Mary S (2006) Characterization of inter-generic hybrids of *Saccharum* using molecular markers. Genet Resour Crop Evol 53:163–169

Oliveira KM, Pinto LR, Marconi TG, Margarido GRA, Pastina MM, Teixeira LHM, Figueira AM, Ulian EC, Garcia AAF, Souza AP (2007) Functional genetic linkage map on ESTmarkers for a sugarcane (*Saccharum* spp.) commercial cross. Mol Breed 20:89–208

Pan YB (2006) Highly polymorphic microsatellite DNA markers for sugarcane germplasm evaluation and variety identity testing. Sugar Tech 8(4):246–256

Pan YB, Burner DM, Legendre BL, Grisham MP, White WH (2004) An assessment of the genetic diversity within a collection of *Saccharum spontaneum* with RAPD PCR. Genet Res Crop Evol 51:895–903

Papini-Terzi FS, Rocha FR, Vencio RZN, Felix JM, Branco DS, Waclawovsky AJ, Bem LEVD, Lembke CG, Costa MDL, Nishiyama MY Jr, Vicentini R, Vincentz MGA, Ulian EC, Menossi M, Souza GM (2009) Sugarcane genes associated with sucrose content. BMC Genomics 10:120

Parida SK, Rajkumar KA, Dalal V, Singh NK, Mohapatra T (2006) Unigene derived microsatellite markers for the cereal genomes. Theor Appl Genet 112:808–817

Parida SK, Kalia SK, Kaul S, Dalal V, Hemaprabha G, Selvi A, Pandit A, Singh A, Gaikwad K, Sharma TR, Srivastava PS, Singh NK, Mohapatra T (2009) Informative genomic microsatellite markers for efficient genotyping applications in sugarcane. Theor Appl Genet 118:327–338

Parida SK, Pandit A, Gaikwad K, Sharma TR, Srivastava PS, Singh NK, Mohapatra T (2010) Functionally relevant microsatellites in sugarcane unigenes. BMC Plant Biol 10:251

Paterson AH (1996) Making genetic maps. In: A.H. Paterson (ed) Genome mapping in plants. R. G. Landes Company, San Diego; Academic Press, Austin, pp 23–39

Paterson AH, Lander ES, Hewitt JD, Peterson S, Lincoln SE, Tanksley SD (1988) Resolution of quantitative traits into Mendalian factors by using a complete linkage map of restriction fragment length polymorphisms. Nature 335:721–726

Pinto LR, Oliveri KM, Ulian EC, Garcia AAF, de Souza AP (2004) Survey in the sugarcane expressed sequence tag database (SUCEST) for simple sequence repeats. Genome 47:795–804

Pinto LR, Oliveira KM, Marconi T, Garcia AAF, Ulian EC, De Souza AP (2006) Characterization of novel sugarcane expressed sequence tag microsatellites and their comparison with genomic SSRs. Plant Breed 125:378–384

Pinto LR, Garcia AAF, Pastina MM, Teixeira LHM, Bressiani JA, Ulian EC, Bidoia MAP, Souza AP (2010) Analysis of genomic and functional RFLP derived markers associated with sucrose content, fiber and yield QTLs in a sugarcane (*Saccharum* spp.) commercial cross. Euphytica 173:313–327

Piperidis N, Jackson PA, Hont AD, Besse P, Hoarau JY, Courtois B, Aitken KS, McIntyre CL (2008) Comparative genetics in sugarcane enables structured map enhancement and validation of marker-trait associations. Mol Breeding 21:233–247

Piperidis G, Piperidis N, D'Hont A (2010) Molecular cytogenetic investigation of chromosome composition and transmission in sugarcane. Mol Genet Genomics 284:65–73

Powell W, Machray GC, Provan J (1996) Polymorphism revealed by simple sequence repeats. Trends Plant Sci 1:215–222

Prasada Rao KSSV, Varier A, Mohapatra T, Kumari A, Sharma SP (2001) Electrophoresis of seed esterases and RAPD analysis for identification of hybrids and parental lines of pearl millet (*Pennisetum glaucum* L.R.Br.). Plant Var Seeds 14:41–52

Prasanna BM (2002) DNA-based markers in plants. In: Prasanna BM (ed) Molecular marker applications in plant breeding. Manual ICAR short-term training course Sep 26–Oct 5, 2002, Division of genetics. IARI, New Delhi, pp 12–17

Price S (1964) Cytological studies in Saccharum and allied genera IX. Further F1 hybrids from *Saccharum officinarum* (2n=80) × *Saccharum spontaneum* (2n–96) Indian. J Sugar Res Dev 8:131–133

Price S (1965) Cytology of *Saccharum robustum* and related sympatric species and natural hybrids. US Dep Agric, Agric Res Serv Tech Bull 1337:47

Price S (1968) Chromosome transmission by *Saccharum robustum* in interspecific crosses. J Hered 59:245–247

Price S (1969) Chromosome numbers in miscellaneous clones of *Saccharum* and allied genera. Proc Int Soc Sugar Cane Technol 13:921–926

Reffay N, Jackson PA, Aitken KS, Hoarau JY, D'Hont A, Besse P, McIntyre CL (2005) Characterisation of genome regions incorporated from an important wild relative into Australian sugarcane. Mol Breeding 15:367–381

Rick CM, Barton DW (1954) Cytological and genetical identification of the primary trisomics of the tomato. Genetics 39:640–666

Rick CM, Dempsey WH, Khush GS (1964) Further studies on the primary trisomics of tomato. Can J Genet Cytol 6:93–108

Riley R, Chapman V, Johnson R (1968) The incorporation of alien disease resistance in wheat by genetic interference with the regulation of meiotic chromosome synapsis. Genetic Res 12:199–219

Roach BT (1989) Origin and improvement of the genetic base of sugarcane. In: Egan BT (ed) In: Proceedings of Australian Society of Sugarcane Technology. Australian Society of Sugarcane Technologists, Queensland, pp 34–47

Rossi M, Araujo PG, Van Sluys MA (2001) Survey of transposable elements in expressed sequence tags (ESTs). Genet Mol Biol 24(1–4):147–154

Rossi M, Araujo PG, Paulet F, Garsmeur O, Dias VM, Chen H, Van-Sluys M-a, D'Hont A (2003) Genome distribution and characterization of EST-derived resistance gen analogs (RGAs) in sugarcane. Mol. Genet Genomics 269:406–419

Rossouw D, Kossmann J, Botha FC, Groenewald JH (2010) Reduced neutral invertase activity in the culm tissues of transgenic sugarcane plants results in a decrease in respiration and sucrose cycling and an increase in sucrose:hexose ratio. Funct Plant Biol 37:22–31

Sacher JA, Hatch MD, Glasziou KT (1963) The sugar accumulation cycle in sugarcane III. Physical and metabolic aspects of cycle in immature storage tissue. Plant Physiol 38:348–354

Salisbury SB, Rose CW (1992) Plant Physiology. Wadswort Publishing Company, California

Santhy V, Mohapatra T, Dadlani M, Sharma SP, Sharma RP (2000) DNA markers for testing distinctness in rice (Oryza sativa L.) varieties. Plant Var Seeds 13:141–148

Sasaki T, Song JY, Kogaban Y, Matsui E, Fang F, Higo H, Nagasaki H, Hori M, Miya M, Murayamakayano E, Takiguchi T, Takasuga A, Niki T, Ishimaru K, Ikeda H, Yamamoto Y, Mukai Y, Ohta I, Miyadera N, Havukkala I, Minobe Y (1994) Towards cataloguing all rice genes-large scale sequencing of randomly chosen rice cDNAs from a callus cDNA library. Plant J 6:615–624

Scott KD, Eggler P, Seaton G, Rossetto M, Ablett EM, Lee LS, Henry RJ (2000) Analysis of SSRs derived from grape ESTs. Theor Appl Genet 100:723–726

Selvi A, Nair NV, Balasundaram N, Mohapatra T (2003) Evaluation of maize microsatellite markers for genetic diversity analysis and fingerprinting in sugarcane. Genome 46(3):394–403

Selvi A, Nair NV, Noyer JL, Singh NK, Balasundaram N, Bansal KC, Koundal KR, Mohapatra T (2005) Genomic constitution and genetic relationship among the tropical and subtropical Indian sugarcane cultivars revealed by AFLP. Crop Sci 45:1750–1757

Selvi A, Nair NV, Noyer JL, Singh NK, Balasundaram N, Bansal KC, Koundal KR, Mohapatra T (2006) AFLP analysis of the phonetic organization and genetic diversity in the sugarcane complex, Saccharum and Erianthus. Genet Resour Crop Evol 53:831–842

Shanthi RM, Alarmelu S, Balakrishnana R (2005) Role of female parent in the inheritance of Brix in early selection stages of sugarcane. Sugar Tech 7(2&3):39–43

Shiringani AL, Frisch M, Friedt W (2010) Genetic mapping of QTLs for sugar-related traits in a RIL population of Sorghum bicolor L. Moench Theor Appl Genet 121:323–326

Singh RK, Srivastava S, Singh SP, Sharma ML, Mohapatra T, Singh NK, Singh SB (2008) Identification of new microsatellite DNA markers for sugar and related traits in sugarcane. Sugar Tech 10(4):156–162

Singh RK, Singh RB, Singh SP, Sharma ML (2011) Identification of sugarcane microsatellites associated to sugar content in sugarcane and transferability to other cereal genomes. Euphytica 182(3):335–354

Sobral BWS, Braga DPV, Lahood ES, Keim P (1994) Phylogenetic analysis of chloroplast restriction enzyme site mutations in the Saccharinaea, Griseb. subtribe of the Andropogonaea Dumort tribe. Theor Appl Genet 87:843–853

Souza GM, Berges H, Bocs S, Casu R, D'Hont A, Ferreira JE, Henry R, Ming R, Potier B, Van Sluys MA, Vincentz M, Paterson A (2011) The sugarcane genome challenge: strategies for sequencing a highly complex genome. Trop Plant Biol doi:10.1007/s12042-011-9079-0

Sreenivasan, TV, Ahloowalia, BS, Heinz DJ 1987 Cytogenetics. In: Heinz DJ (ed) Sugarcane improvement through breeding. Elseveir, Netherlands, pp 211–254

Srivastava BL, Cooper M, Mullins RT (1994) Quantitative analysis of the effect of selection history on sugar yield adaptation of sugarcane clones. Theor Appl Genet 87:627–640

Stevenson GC (1965) Genetics and breeding of sugarcane. Longmans Green and Co Ltd., London, p 282

Sugiharto B, Sakakibara H, Sumadi, Sugiyami T (1997) Differential expression of two genes for sucrose-posphate synthase in sugarcane: molecular cloning of the cDNAs and comparative analysis of gene expression. Plant Cell Physiol 38(8):961–965

Swapna M, Sivaraju K, Sharma RK, Singh NK, Mohapatra T (2011a) Single-strand conformational polymorphism of EST-SSRs: a potential tool for diversity analysis and varietal identification in sugarcane. Plant Mol Biol Rep (2011) 29:505–513

Swapna M, Srivastava S, PandeyDK (2011b) Targeting genes linked to sugar content for quality improvement: A molecular marker approach in sugarcane. Proceedings of X Agricultural Science Congress, 8–10 Feb. 2011, NBFGR, Lucknow. p 214

Taiz L, Zeiger E (1998) Plant physiology. Sinauer Associates Inc. Publishers, Sunderland, pp 214–215

Tanksley S (1993) Mapping polygenes. Ann Rev Genet 27:205–233

Tsuchiya T (1959) Genetic studies in trisomic barley.I. Relationships between trisomics and genetic linkage group of barley. Jap J Bot 17:177–213

Tsuchiya T (1991) Chromosome mapping by means of aneuploid analysis in barley. In: Gupta PK, Tsuchiya T (eds) Chromosome engineering in plants: genetics, breeding and evolution part A. Elsevier, Amsterdam, pp 361–384

UPOV (1992) UPOV International convention for the protection of new varieties of plants, UPOV publication No. 221(E). International union for the protection of new varieties of plants, Geneva

van der Merwe MJ, Groenewald JH, Stitt M, Kossmann J, Botha FC (2010) Down regulation of pyrophosphate:D-fructose-6-phosphate 1-phosphotransferase activity in sugarcane culms enhances sucrose accumulation due to elevated hexose-phosphate levels. Planta 231:595–608

Vettore AL, da Silva FR, Kemper EL, Arruda P (2001) The libraries that made SUCEST. Genet Mol Biol 24(1–4):1–7

Vicentini R, de Maria Felix J, Dornelas MC, Menossi M (2009) Characterization of a sugarcane (*Saccharum* spp.) gene homolog to the brassinosteroid insensitive-1-associated receptor kinase I that is associated to sugar content. Plant Cell Rep 28:481–491

Vos P, Hogers R, Bleeker M, Reijans M, van de Lee T, Hornes M, Frijters A, Pot J, Peleman J, Kuiper, MZ (1995) AFLP: a new technique for DNA fingerprinting. Nucleic Acids Res. 18:6531–6535

Wang X, Bowen B (1998) A progress report on corn genome projects at pioneer hi-bred. In: Plant and animal genome conference VI. San Diego, California

Wang J, Roe B, Macmil S, Murray J, Tang H, Nafar F, Wiley G, Bowers JE, Chen C, Rokhsar DS, Hudson ME, Moose SP, Paterson AH, Ming R (2010) Microcollinearity between diploid sorghum and autopolyploid sugarcane genomes. BMC Genomics 11:261

Whitkus R, Doebley J, Lee M (1992) Comparative genome mapping of sorghum and maize. Genetics 132:1119–1130

Whittaker A, Botha FC (1999) Pyrophosphate:D-fructose-6-phosphate 1-phosphotransferase activity patterns in relation to sucrose storage across sugarcane varieties. Physiologia Plantarum 107:379–386

Williams JGK, Kubelik AR, Livak KJ, Tingey JA, Rafalski SV (1990) DNA polymorphism amplified by arbitrary primers are useful as genetic markers. Nucleic Acids Res 18:6531–6535

Williams LE, Lemoine R, Sauer N (2000) Sugar transporters in higher plants: a diversity of roles and complex regulation. Trends Plant Sci 5(7):283–290

Wu l, Birch RG (2007) Doubled sugar content in sugarcane plants modified to produce a sucrose isomer. Plant Biotechnol J 5:109–117

Wu KK, Burnquist W, Sorrells ME, Tew TL, Moore PH, Tanksley SD (1992) The detection and estimation of linkage in polyploids using single-dose restriction fragments. Theor Appl Genet 83:294–300

Yamamoto K, Sasaki T (1997) Large-scale EST sequencing in rice. Plant Mol Biol 35:135–144

Yap EPH, McGee JO'D (1992) Nonisotopic SSCP detection in PCR products by ethidium bromide staining. Trends Genet 8:49

Zanca AS, Vicentini R, Ortiz-Morea, Bern LED, da Silva MJ, Vincentz M, Nogueira FTS (2010) Identification and expression of microRNAs and targets in the biofuel crop sugarcane. BMC Plant Biol 10:260